If I Could Tell You Another Thing

A further collection of advice and pedagogical insights from mathematics teachers

Edited by David Miles

THE MATHEMATICAL ASSOCIATION

If I Could Tell You Another Thing

© 2023 The Mathematical Association

This book is dedicated to the selfless volunteers of The Mathematical Association, past, present and future, especially the contributors to this book who have so generously offered their valuable time and expertise. A special acknowledgement goes to Chris Pritchard whose skilful typesetting, insightful counsel and unwavering patience were instrumental in bringing this project to fruition.

CONTENTS

1	Introduction DAVID MILES	1
2	Teach pupils from where they are, not where we want them to be DAVE TAYLOR	3
3	The importance of a thorough prerequisite knowledge check AMANDA AUSTIN	7
4	It's well worth the wait CHRIS PRITCHARD	14
5	Give resit students space to explore mathematics REBECCA ATHERFOLD	16
6	Using educational books to develop your pedagogy RHIANNON RAINBOW and DAVID TUSHINGHAM	20
7	Teaching ratio and proportion DAWN DENYER	24
8	Choosing the right tasks for your A level mathematics lessons SUSAN WHITEHOUSE	32
9	Invest time and thought into your demonstrated examples NIKKI ROHLFING	35
10	Teaching problem solving SHEENA FLOWERS	38
11	Planning and teaching for coherence KATHRYN DARWIN	44
12	Mathematics is important for its own sake MARK DAWES	51
13	Noticing and wondering STELLA DUDZIC	57

14	Don't stop interweavin' NATHAN DAY	62
15	Thoughts on A level integration TOM BENNISON	72
16	When am I ever going to use this? DAVE GALE	81
17	I wouldn't tell you anything CHARLOTTE HAWTHORNE	86
18	Expect maths teachers to agree on 'good learning' but not 'good teaching' JEN SHEARMAN	90
19	Go off-piste PETER RANSOM	96
20	Develop your questioning skills TOM BUTTON	103
21	Take time to explain why DAVID MILES	107
22	Better formative assessment review DARREN CARTER	111
23	Numeracy and the importance of maths teachers SUSAN OKEREKE	116
24	Examiners are your friends GRAHAM CUMMING	121
25	Using whiteboards in the mathematics classroom ROB SOUTHERN	124
26	Good enough MEL MULDOWNEY	129
	Contributors	133

1   Introduction

David Miles

The Mathematical Association was established in 1871 as the Association for the Improvement of Geometrical Teaching and is believed to be the oldest subject association in the world. For more than one hundred and fifty years, it has been a prominent and well-informed participant in the ongoing conversation and discourse on the most effective ways to teach and learn mathematics. By fostering a community of professionals committed to ongoing evidence-based progress, the Association has played a significant role in the advancement of mathematics education in the United Kingdom. It has never wavered in its determination to support the professional development of its members, offering a wealth of resources, publications and journals as well as events, conferences and networking opportunities that serve as a catalyst for continued learning and improvement.

I became actively involved with The Mathematical Association fifteen years ago as an occasional contributor to their secondary journal Mathematics in School and this soon led to an invitation to join Teaching Committee. For the past decade, I have served on the Council of the Association in a variety of roles and have recently had the privilege of returning to Teaching Committee as Chair. My full-time job as a classroom teacher and senior leader, coupled with the aforementioned volunteer work, has given me a deep appreciation of the daily challenges faced by secondary mathematics teachers. In the process of editing this book, I have drawn upon much of this knowledge and experience.

I have thoroughly enjoyed compiling *If I Could Tell You Another Thing* over the past few months and am excited to present this new collection of expert guidance and pedagogical reflections. The previous volume, *If I Could Tell You One Thing*, was the brainchild of Ed Southall and was so popular and widely acclaimed that it was natural to immediately commission a follow-up. For this book, I decided to expand the diversity of views by approaching a completely different set of authors and it has been a pleasure to collaborate with so many visionary, dedicated and inspirational colleagues. Our contributors may be at a variety of points in their careers but they all have impressive profiles and have earned the respect of the profession for their common sense, expertise and commitment to student success. Their pragmatic advice, rooted in skilful practice, will provide readers with valuable strategies, techniques and perspectives they can seamlessly implement in their own classroom to enhance the teaching and learning of mathematics.

The *If I Could Tell You ...* series has an obvious audience amongst early career teachers but the positive feedback we have received makes it clear they have a much broader appeal. The content is informative, wise and thought-provoking and every chapter is infused with a contagious enthusiasm that has the capacity to inspire and reinvigorate the weariest and most disillusioned of colleagues. These books are not only tools for learning but also a stimulus for mathematics teachers to continue to grow and excel in their careers. School leaders, teacher educators, and any individual involved in the schooling of young mathematicians will find plenty to savour within these pages.

It is my hope that *If I Could Tell You Another Thing* will prove to be equally as influential and impactful as its predecessor, serving as a high quality resource for any secondary school teacher of mathematics and providing inspiration and practical, actionable guidance on how to effectively deliver key aspects of our challenging but beautiful subject.

2   *Teach pupils from where they are, not where we want them to be*

Dave Taylor

"Pupils don't learn from what you say and do." When I first heard this, I didn't believe it. "Of course they do... I describe, explain, prompt, show, and this results in pupils learning. How else would it go?" I thought. The addition to "Pupils don't learn from what you say and do" that made this make sense to me was "they learn from their interpretation of what you say and do". After more listening, a bit of reading and some mental wrestling, I succumbed to the statement that "pupils don't learn from what you say and do".

In the 2000 'comedy' film Road Trip, Josh Parker (played by Breckin Meyer) needs to get back from his cross-country trip for a midterm exam in ancient philosophy, but knows nothing of the content. This is a concern, because if he doesn't pass the exam he'll be kicked out of school. His friend Rubin (played by Paulo Costanzo) says he can teach him ancient philosophy in 46 hours, claiming, "I can teach Japanese to a monkey in 46 hours. The key's just finding a way to relate to the material." Relating ancient philosophy to the 1990s phenomenon that was professional wrestling, he successfully (scoring a B+, as required to pass the class) teaches him the content as they travel across the country.

When a teacher describes, explains and shows, pupils are continually making links between the new knowledge and their already-constructed schemata. They draw on their individual prior experiences and well-embedded knowledge to make meaning with the new content. Breckin Meyer's character was a fan of professional wrestling, and having Socrates described as the Vince McMahon of philosophy allowed him to make meaning with content that he was previously struggling with.

Throughout the early stages of my 15 years in the classroom, I was guilty of not giving this notion as much attention as I should have. As an inexperienced teacher I dutifully followed the scheme of learning set out by my head of maths, and as an example, went ahead with planning a lesson on 'solving equations with the unknown on both sides' for my year 9 bottom set class. The way that I went about this, and then teaching the lesson, had been successful with other classes, but wasn't with this class. I would put it down to their being poor learners, that I'd done all that was within my power, but the professional development I have since undertaken has taught me that this was simply not the case – all pupils can learn mathematics well.

The issues that I experienced were a result of the conveyor belt model of education – one which considers teaching pupils in the same way as manufacturing cars. The chassis is loaded on to the conveyor belt, and machines along the conveyor belt do their jobs, adding the side frames, doors, hoods and roofs to the chassis, applying the paint job, installing the engine, and so on. Sometimes the conveyor belt malfunctions, and the task isn't completed as it was supposed to, resulting in a faulty product. If we treat teaching in this way, we'll ultimately end up with a faulty product argued as a 'poor learner', but it doesn't have to be this way. Pupils shouldn't move on to 'solving equations with the unknown on both sides' if they haven't developed a robust and flexible understanding of 'solving two-step equations', and an argument shouldn't be made for them to move on to more complex material simply because the Earth has orbited the Sun one more time, rather than whether they are ready for the new material or not.

If we teach pupils from where they are, building on prior knowledge, rather than teaching the content from the mathematical level we want them to be, every pupil can learn well. In the example of solving equations, pupils in my year 9 class were able to solve one-step and two-step equations, but their method was to use trial and improvement to successfully find the value of the unknown, rather than inverse operations. What this meant for my lesson, was that teaching them using the balance method when their understanding of inverse operations was almost non-existent was unlikely to result in any learning. Pupils were unable to form connections with prior knowledge, and they saw 'solving equations with the unknown on both sides' as a completely unrelated skill to 'solving two-step equations'.

Our schemes of learning should be constructed to respond to pupils' expertise, rather than their ages. We must respect the knowledge that pupils have, as well as the gaps that exist in their mathematical universe, to ensure that our lowest-attaining pupils are not being taught material that is too complex for them and that our highest-attaining pupils are not having their mathematical development hindered by a conveyor belt that is moving too slowly for their curiosity. A simplistic model to use as a scheme of learning would be to write a hierarchical list of skills for solving equations, as an example, and avoid attaching age-related expectations to each skill. This will allow teachers to identify the appropriate mathematics for the pupils in front of them.

This approach will enable all pupils to be successful in mathematics, building on their prior knowledge and banishing the myth that 'maths is just for smart kids'. Pupils will experience a shift in mindset from 'I hate maths' to 'I CAN do maths', developing motivation through their successes in every lesson and feeding the virtuous cycle between success and motivation.

The most important pedagogy is a responsive one, and in the aforementioned simplistic model, the teacher can assess pupils' expertise, identifying the level of their understanding in solving equations, and develop a bespoke learning pathway for the pupils in their care, ensuring that the complexity of the mathematical ideas being taught are appropriately pitched to bring about success. Where pupils encounter a stumbling block, the teacher should abandon the plan, reassess prior knowledge, mitigate against the identified misconceptions and alter their explanations, examples and models to overcome these hurdles, resulting in strong connections between pupils' existing schemata and the new learning.

If it's vital that we acknowledge pupils' prior experiences, and that pupils learn from their interpretation of what we say and do, we should also consider the variety of tasks that we provide pupils with in order to build up a rich bank of experiences. The best way for us to do this in the mathematics classroom is to explain concepts using multiple representations through the use of concrete manipulatives such as two-colour counters and algebra tiles. We can then make use of diagrammatical equivalences so that pupils have multiple strategies beyond the four walls of the classroom and are not limited by the physical resources of their current location, and ultimately phase out the use of concrete manipulatives and pictorial representations as expertise develops in the abstract.

Following the example of my year 9 pupils working on solving equations, pupils must first appreciate the idea of doing and undoing with inverse operations and the concept of 'equalness' and two expressions being 'level', before being exposed to these ideas being used together with a physical model of weighing scales, to stress the importance of 'doing the same to both sides' whilst applying the 'balance method' for solving equations. This method would be preferable to trial and improvement, because we are intending to use this idea to progress their learning further as expertise grows, and we can begin to develop the use of abstract notation with 'solving one-step equations'. We must insist that pupils write their workings so that the use of the balance method becomes more familiar, and not accept only solutions from pupils as their written work, as this is to lessen the importance of the balance method.

As we move on to 'solving two-step equations', pupils may be enticed by the success of trial and improvement as a method of bringing about a solution, but having developed understanding of the balance method when working with less-complex material, we can make use of physical scales as a concrete manipulative and introduce algebra tiles to experience the solution steps in a different way. Whilst using the algebra tiles, we can begin to represent the solution steps pictorially, and whilst using diagrams, we can phase into the use of abstract notation, building on what has come before. Once pupils have developed robust and flexible learning with solving these more complex equations, we can move on to 'solving equations with the unknown on both sides' making use of their prior experiences, extending the process by adding in the simple step of isolating the unknown on one side of the equals sign.

Applying the solution steps when 'solving equations with the unknown on both sides' without this rich depth of understanding, matured over time, is unlikely to lead to success with this new learning. Early career Dave would have given little thought to this progression through a mathematical pathway, but present-day Dave is more aware. We must teach pupils from where they are, not where we want them to be.

3   *The Importance of a Thorough Prerequisite Knowledge Check*

Amanda Austin

A while ago, I sat down to watch season 4 of popular BBC drama *Killing Eve*. I had enjoyed seasons 1 to 3 and was excited to see what the new season had to offer. So, I tuned in, watched the short recap that came at the beginning of the new series, and made my way through episode 1, and then episode 2. However, much as I tried, I just couldn't get into this new season. Even though I was familiar with all the main characters and had a reasonable recall of the plot from the previous series, things weren't quite making sense. It was all quite complicated and there were too many gaps in my knowledge, too many small details from previous seasons that I'd forgotten all about. I abandoned season 4, disappointed and frustrated.

Similarly, each time we introduce a new 'season' of mathematics, it is vital that we ensure that students have a thorough grasp of what has come before – not just the 'main characters' but the small details and nuances too. Only then can we enable them to make sense of new ideas and content, and hence gain maximum enjoyment and understanding of the mathematics being presented to them. In other words, prior to teaching any new topic, we must carry out a thorough prerequisite knowledge check, using the information gained from this check to identify gaps in knowledge and understanding and inform the starting point for our teaching of any new content.

So, exactly what does a **thorough** prerequisite knowledge check look like? Well, let's start by imagining I'm going to be introducing my class to repeated percentage change using a calculator. In the past, I would have wanted to check two things:

a) Can a student write a multiplier for a percentage increase or decrease?

b) Can a student carry out a percentage increase or decrease using this multiplier method on their calculator?

To do this I would probably have incorporated it into a retrieval/recall starter that looked something like this, and would be carried out by students on mini-whiteboards:

LAST LESSON	LAST TOPIC
A car travels at 40 km/h for 1.5 hours. How far has it travelled?	Write down the gradient of the line with equation $y = -3x + 7$

LAST TERM	LAST YEAR
Expand and simplify $2(x + 5) + 3(x - 1)$	(a) Write down the multiplier for a 25% increase. (b) Increase $75 by 25%

Looking at this now, I think I can safely say it is most certainly **not** a thorough prerequisite knowledge check. For a start, I've only checked whether or not students can write the multiplier for a percentage increase and a simple one at that. What about the multiplier for a 13.5% decrease? And do students understand how the multiplier is formed, or are they just recalling a commonly-used multiplier for a common percentage increase? Secondly, if they get the correct answer to part b, having successfully increased $75 by 25%, have they actually recalled and used the multiplier method? Finding 25% of a number is a calculation that some students could do in their heads. I've not really picked a question that forces them to use the multiplier method and their calculator.

So, let's improve on this. How about still checking the same two prerequisites but through a series of questions that get increasingly more difficult and address some of the concerns about the previous check. Something like this, again carried out by students on mini-whiteboards:

QUESTION 1	QUESTION 2
(a) Write down the multiplier for a 15% increase. (b) Increase £75 by 15%	(a) Write down the multiplier for a 20% decrease. (b) Decrease £244 by 20%

QUESTION 3	QUESTION 4
(a) Write down the multiplier for a 43% increase. (b) Increase £13 by 43%	(a) Write down the multiplier for a 7.5% decrease. (b) Decrease £96 by 7.5%

This is definitely an improvement. This time I'm checking both increase and decrease, as well as throwing in some more challenging percentages. But it's still only checking procedural knowledge, and not really delving any deeper. If I'm aiming to assess students' prior knowledge and understanding more thoroughly then I really should be extending this further – check if students can **explain** how to find a multiplier, spot **common misconceptions** and, looking forward to carrying out repeated multiplication, **understand** that this can be written in power notation. This way, rather than dealing with these issues as and when they crop up during a sequence of lessons on the new topic, I can assess whether they do indeed need addressing and therefore plan for it in advance. This leads me to my most recent iteration of a prerequisite knowledge check and one that I hope fits the brief of being thorough. For repeated percentage change, it might look something like this:

QUESTION 1

A student wants to increase £36 by 17% and **correctly** calculates
$$£36 \times 1.17 = £42.12.$$

(a) Explain why the student has multiplied by 1.17.

(b) What calculation would they do to increase £36 by 47%?

QUESTION 2

There are 450 bees in a hive. The number of bees increases by 8%. How many bees are now in the hive?

A student answers this question **incorrectly** using this calculation:
$$450 \times 1.8 = 810 \; bees.$$

(a) What mistake has the student made?

(b) Calculate the correct answer to the question.

QUESTION 3

(a) (i) Write down the multiplier for a decrease of 15%.

  (ii) Hence decrease £75 by 15%.

(b) Decrease £75 by 8.5%

(c) Decrease £75 by 91%

QUESTION 4

(a) A lamp which costs £29.95 is reduced in price in a 35% off sale. What is the sale price of the lamp?

(b) A rapidly-increasing rabbit population increases by 185% in one year. If there were originally 236 rabbits, how many are there after a year?

QUESTION 5

(a) Write $2 \times 2 \times 2 \times 2 \times 2$ in power notation.

(b) Write $0.85 \times 0.85 \times 0.85$ as a single power of 0.85.

QUESTION 6

Using your calculator, evaluate:
(a) 4.5×2.6^3

(b) $300 \times 1.12^5 \times 0.9^2$

This check is a lot more detailed and is going to provide me with much more information on my students' understanding of the prerequisites required for teaching repeated percentage change. I think it helps to explain my thinking in using each of the questions:

- In Question 1, I have focused on multipliers for a percentage increase. Can students **explain** how a multiplier of 1.17 represents a percentage increase of 17%? Can they find the multiplier for a different percentage increase?

- In Question 2, I am addressing one of the common misconceptions for multipliers, checking to see whether students can spot that the multiplier for an 8% increase is not 1.8 but 1.08, and then giving them the opportunity to correctly answer this question.

- In Question 3, I am checking in detail whether students can use their calculators to evaluate amounts involving powers and, if they can find a reasonable straightforward multiplier for a 15% decrease, can they also find multipliers for more challenging percentage decreases?

- In Question 4, I am introducing some worded problems, to check whether students are confident converting these into a percentage change calculation. This question also has the bonus of checking about rounding – will they remember to round money to 2 decimal places and what will they do when faced with 672.6 rabbits? Additionally, there is a tricky 185% percentage increase and I want to know if students can confidently calculate this more challenging multiplier.

- In Question 5, I am checking another prerequisite for this topic – are students able to write a repeated multiplication as a power, with both integers and decimals. If they can't, I would like to know this before I launch into this new idea.

- Finally, in Question 6 I am checking that students can use their calculators to evaluate calculations involving powers.

I hope you agree that this more detailed, thorough prerequisite knowledge check will not only give me more information, but also very specific information about what my students know and don't know, and whether they are ready to begin learning about repeated percentage change. Clearly, this check is not something that can be incorporated into a quick starter or rushed through by students in the last five minutes of a lesson. It's going to take up a sizeable chunk of lesson time; however, in my experience this time will be recouped in the long run. Additionally, the level of detail of the check means it probably won't work well using mini-whiteboards, which would be my usual method of checking for student understanding. Whilst no doubt the chosen method of carrying out this detailed prerequisite check will vary according to factors such as the topic or class being taught, of the different ways to implement this, my experience is that printing A3 versions and having students work in pairs has been most successful for the classes I teach. Alternatively, you may choose to give your knowledge check as a mini-quiz, where students answering independently allows you to pinpoint exactly what prior knowledge each individual student has. Another option, particularly if lesson time is at a premium, could be to set the prerequisite knowledge check as a piece of homework.

Once students have completed their prerequisite knowledge checks, my preference is to collect them in, review them and then use the information gathered to inform planning. More experienced teachers may choose to use the knowledge check at the beginning of a lesson, then once completed, work through each of the questions and use them as a basis to promote discussion, modelling solutions, tackling knowledge gaps and misconceptions head-on, and adapting their teaching accordingly.

Whichever method you choose to implement the more thorough prerequisite knowledge check, a good one should allow the teacher to:

a) Identify which prior knowledge is secure in all, most or only a few students, and hence whether anything from a quick recap to a full re-teach is required

b) Identify whether further independent practice of any of the prerequisite skills is required to consolidate understanding and increase student confidence

c) Identify any misconceptions which should be addressed prior to introducing new content and ideas

d) Pinpoint the best place to start the new topic in order that all students can confidently build on what they already know and access new content at a level appropriate to them.

A further example of a thorough prerequisite knowledge check for circle theorems is presented below. Whilst the time taken to develop, carry out and analyse these prerequisite knowledge checks is certainly not insignificant, my experience is that, by implementing them, the subsequent sequence of lessons is more successful for both teacher and students, with the latter demonstrating an increased confidence when the intended new idea or concept is eventually introduced.

QUESTION 1

Match the part of the circle with its name:

Diameter: _____

Radius: _____

Tangent: _____

Chord: _____

QUESTION 2

(a) Work out the value of x. Give a reason for your answer.

(b) Work out the value of y. Give a reason for your answer.

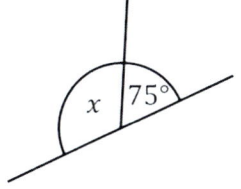
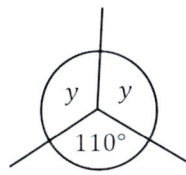

QUESTION 3

(a) Write down the name of the type of triangle shown.

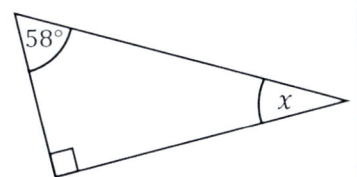

(b) Work out the size of angle x. Give a reason for your answer.

QUESTION 4

Work out the value of angle x, giving reasons for your answer.

A student answers this question **incorrectly**, writing:

$x = 56°$ because base angles in an isosceles triangle are equal

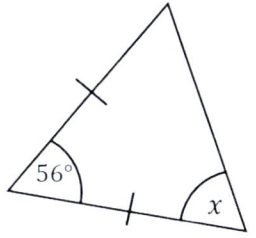

(a) What mistake has the student made?

(b) Answer the question correctly.

QUESTION 5

(a) Write down the size of angle CAB.

(b) Write down the size of angle BCD.

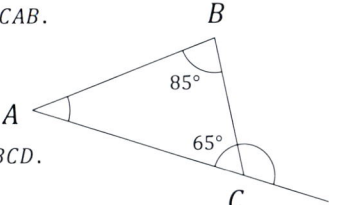

4   It's Well Worth the Wait

Chris Pritchard

It was only in the latter part of my career that I became aware of something called the 'wait time' and, as I experimented with it in my lessons, I discovered just how effective it is.

There were always aspects of teaching that I thought I was rather good at almost from the outset and they were the same ones I continued to develop. I could explain what was going on, how things worked, what followed from what, I could normally conjure up a visual image to support understanding and I could ask suitable, and often searching, questions. But I suspect that like many mathematics teachers, when I did ask a question of a particular pupil as part of our whole-class deliberations, I would quickly move on to someone else if an appropriate response was not forthcoming within a short space of time. I was missing a trick!

If I recall correctly, it was Dylan Wiliam, *Inside the Black Box* and *Assessment for Learning* that first prompted me to look into the wait time issue. Wiliam seemed to think that everyone was aware of the original and subsequent research, so I initially felt somewhat uncomfortable that I didn't. In fact it's exactly fifty years since the idea of extending the wait time was raised by Mary Budd Rowe. Where many teachers waited no longer than a second and a half before jumping in and perhaps answering the question themselves or else turning to another member of the class, Rowe and her successor researchers became aware that a slew of beneficial pedagogical outcomes followed if teachers simply held fire and extended the wait time to three seconds or more. The standard "I don't know" response largely disappeared to be replaced by longer, more detailed and more accurate responses, and assessment scores increased.

I remember the first time I used the new tack, not the exact problem we were considering but the sort of things I was saying and the sort of responses I was getting:

    Daniel, can you help with this one?

    No, I don't think so.

    Take your time, there's no hurry.

> *… No, I don't know what to do.*

I could have given up at this point. There were three or four hands up and someone called out "I know". But I stayed with Daniel.

> What have we done before when we've been stuck on a question?
>
> *… Checked out what information we've been given?*
>
> Ok, so what have we been given?
>
> *The height of a building and an angle.*
>
> And what do we need to find?
>
> *A distance along the ground. I've got it now, it's trigonometry and we need to use tangent.*

There was an English teacher in my first school, a large comprehensive in the East End of London, who didn't just speak quietly to his classes, he literally spoke in a whisper. The school building had been extended over the years and some parts were open plan. I had to tiptoe along the edge of his teaching area so as to not disturb his class, so I witnessed his approach at first hand. I thought at the time that it was completely weird. But … and it's a big BUT, the class hung on his every word. Now with my extended wait time I found the same thing happening. Every member of the class knew that I was not going to accept an "I don't know" response, so they were alert to the possibility that they would be the next to be asked and they worked much harder to have an answer ready. In moving from one and a half seconds to three seconds, everyone in the room was on task, focussed and ready to contribute.

So, if I could tell you just one thing it would be to take a little more time when questioning. It's well worth the wait.

5   *Give Resit Students Space to Explore Mathematics*

Rebecca Atherfold

I had over 15 years' experience in schools before I moved into the FE sector. I thought I had a reasonably good idea of what to expect. I had worked with children with a wide range of attainment and I knew how horribly low the resit pass rate was. Yet I was shocked many times in my first few weeks. Even now, as a seasoned GCSE resit teacher, I am frequently surprised. This reaction is good. It should always be shocking that a young person can leave school after 12 years of education and not have any method for subtraction or be able to tell the time.

Of course, this isn't the whole picture. Most resit students have learnt a lot of maths in their years at school. It is true that some basic skills are often lacking but that's not to say we need to start everything from scratch.

So, where do we start?

The temptation is to start at the end, with the exam that the students are due to take. Whether they are studying Functional Skills or GCSE, all FE maths students have an exam (or three) coming up. Look at the specification, see what's on it, chunk it up into lessons – scheme of learning written. Except, in the case of GCSE, that gives us 8 months to teach an entire key stage. However you interpret the term 'mastery', this is the complete opposite. Even with the smaller specification of Functional Skills, this approach has the potential to lack the depth required to be able to answer questions that draw on more than one area of mathematics. 75% of the Functional Skills exam is problem solving, so depth is crucial for success.

This approach also ignores the students themselves. Nobody would argue that teaching calculating using standard form should come before we teach integer calculation or place value. Yet, it is not uncommon in FE to see the former being taught to students who have at best a shaky grasp of the latter because we assume that they should know it. It is not surprising that the concepts being taught don't stick, even though the scheme of learning is being covered.

We know resit students will have gaps. We need to find them so we can ensure the students have the necessary prerequisites for whatever we are planning to teach. We also know that resit students are not blank slates. Most students will, at the very least, have been exposed to the topics on a GCSE resit scheme of learning before. Alongside the gaps, there will be secure learning. Our job is to find it and build on it.

Again, the temptation is to start at the end and use a past paper, or past exam questions to diagnostically assess. I am unconvinced of the value of this, even with the most straightforward AO1 questions. For example, a student may answer:

$$m + m + m + m = 4m.$$

Great! They can collect like terms. But do I really know that? Do I know that they weren't deliberating between $4m$ and m^4 and just happened to pick the right one?

A question I am asked a lot is, "Is this the one where you …?" Resit students have been around the block and know if they see a rectangle, they will probably have to add the sides or multiply two of them. They've grown to see it as almost arbitrary.

In addition, we have only have a couple of lessons to demonstrate that maths at college can be different from maths at school. Do we want to launch straight in with the very thing that caused them to "fail" at the end of Year 11?

Good diagnostic assessment is challenging in any setting. In FE it is particularly hard. 16 year olds arrive at college nervous and vulnerable and determined to look anything but. They are trying to work out who they are and how they want to appear to their new peers. They are experienced in how to survive maths classes and know how to fly as far under the radar as possible. Alternatively, they might be very proficient in using behaviour so that maths gets avoided altogether. But we must get this right to dodge the temptation to start at the end of the course and instead meet the students where they are right now.

My approach to this has evolved over the past few years. I have long been interested in introducing manipulatives in my FE classrooms. When I taught in primary, they were a vital part of my and my pupils' toolkits. As part of the Centre for Excellence in Maths (CfEM) project (see Note), I led action research into using manipulatives in FE, which enabled me to explore their use in this setting more thoroughly. The outcomes were positive for both teachers and students. But what really stood out for me was their use as a diagnostic tool. For example, I asked students to show me a half on a hundred bead string.

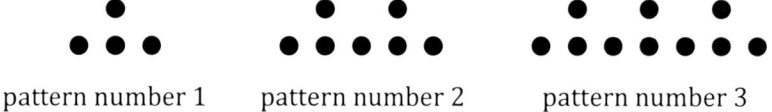

I noticed that some learners were counting 50 beads one by one, even though the groups of 10 are easily identifiable, and they had no problem identifying that half of 100 is 50. Not being to see the importance of 10 in our number system, the 'ten-ness' of ten, is a huge gap and has implications for so many areas of maths. It is also something that I don't think any written assessment designed for GCSE students would have shown me because we assume 16 year olds can count in multiples of 10. Suddenly, I had an insight as to why they were finding rounding so difficult. I had to change what I had planned to teach and go back a few stages. Time well spent.

Now I plan to use manipulatives diagnostically. Recently, I was teaching sequences to a class where most students achieved a grade 3 last year.

Here is a sequence of patterns made with counters.

pattern number 1   pattern number 2   pattern number 3

Find an expression, in terms of n, for the number of counters in pattern number n.

Experience tells me that, when faced with a question like this, most resit students will count the difference between the patterns and, in this case, write one of $3n$, $n + 3$ or $n = 3$. They might be confusing term to term with position to term, or spotting the 3, putting an n in somewhere and hoping for the best.

I started by giving my class counters and some nth term expressions to represent using the counters. Then they created sequences and found expressions, in terms of n. For example, $2n + 1$ can be represented as:

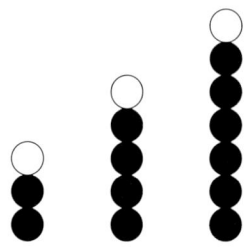

Representations allowed the students to see the difference between $2n$ and $n + 2$. However, great value also came from the discussions they had with each other, and with me. I answered questions about the difference between an expression and an equation, why there is no equals sign in an nth term expression, why it is 'n' not 'x'. I heard a student explain to their partner that the "nth term is just a maths word for the rule of the sequence". I could see which students understood the principle but were let down by their knowledge of working with negative numbers. Students who were confident with linear sequences tackled quadratic ones with the counters. They were engaged, which isn't something that can be taken for granted in a resit class. I was excited by what I was seeing, the students could see that and were proud of their work. Again, not something in FE that can be taken for granted!

I could talk about the power of manipulatives all day. I had a lightbulb moment this morning when I introduced double-sided counters to a class. They initially struggled with the concept of zero pairs because they did not know that $-6 + 6 = 0$. It is easy to assume 16 and 17 year olds know this and maybe this explains why some students in this class struggle to solve equations?

This is why I believe that manipulatives have a special place in FE. Not just because of the value of representations but because they give students the space to revisit maths they have seen before, show what they know and engage with it – dare I say, even enjoy it.

So, if I could go back and tell my younger self one thing when I started working in FE, it would be to plan to give the students space to explore as much as possible in every lesson. They will learn from it. You will learn what you need to teach them, where they have the fundamental gaps that are stopping them from succeeding. Resist temptation to focus solely on the exam and total coverage of the specification. Nobody lost out on a Grade 4 because they couldn't remember an exact trig value.

Note

Centres for Excellence in Maths (CfEM) was a five-year national improvement programme, funded by the DfE, aimed at delivering sustained improvements in maths outcomes for 16 to 19 year olds, up to Level 2, in post-16 settings.

6 Using Educational Books to Develop your Pedagogy

Rhiannon Rainbow and David Tushingham

We have all been there. The night before, lost in our planning thoughts. Pondering how to get our students to the end goal for the next sequence we are teaching, unsure of which task we should use. There's limited time. We have some ideas, but it just doesn't seem to flow. It's late; there's no one to ask. Resources and websites are inspiring emerging ideas, but something is missing. We need to make a decision. We need to put the lesson together now, but we continue to procrastinate. We need advice.

If we could tell you one thing, it would be the power of a good pedagogical book. There are so many excellent maths-specialist books that have been written, each offering 'the lightbulb' for constructing, evolving and applying our pedagogical practice. The books share expertise on mathematical tasks, how to teach maths through examples, and visible mathematics for enhancing your subject knowledge. There is a book out there for just about every problem that a maths teacher may encounter when planning and teaching a high-quality learning sequence. The common paradox we find ourselves in is that the more we learn, the less we feel we actually know. However, when it comes to the teaching of mathematics it is as simple as, the more you know about it, the more you really do know. We would go even further to say that the more you read, the more links you can make between what you read, and the more ideas you have to take away and try in the classroom. As a result, you become more expert as a teacher.

We all want to improve, but finding the time to read is difficult. Joining an educational book club can really help to navigate the wealth of advice and expertise that is available, making it easier to locate the advice you are looking for. We were aware of the demand for a resource that signposts educational texts and we had the time to build such a resource. In co-founding the GLT book club, we created a platform where teachers could connect directly with the authors. This was often framed around considering a key concept, barrier or misconception for the teaching of our subject. At the end of each session, the take-away would 'tell you one thing' about the key messages and how it could translate to the classroom. Our book club is designed to dip in and out of, so that you can quickly find that inspiration when solving the 'night before planning problem'.

The authors we have spoken to have offered the highest quality of scaffold for our reflections. The recordings and linked resources for these podcasts are available here, should they be useful too. The books we chose to read and discuss were an incredible support for us and we highly recommend them!

Sign up to the GLT Always Learning Network for free to access the GLT & Friends Book Club recordings and supporting resources using the links below:

Sign up
https://bit.ly/GLTalwayslearning

GLT & Friends Book Club Community
https://bit.ly/GLTBookClub

If we could tell you one thing from each of the first ten maths-specific books we have looked at so far, they would be:

(1) **Mark McCourt**, 2019. *Teaching for Mastery*, Woodbridge: John Catt.

When students engage in problem solving, use problems where the knowledge required is prior knowledge. This means that students will have to think less about the mathematical concepts and can instead immerse themselves in how to link mathematical knowledge in order to solve the problem.

(2) **Peter Mattock**, 2019. *Visible Maths: Using Representations & Structure to Enhance Mathematics Teaching in Schools*, Carmarthen: Crown House Publishing Limited.

Manipulatives can be a powerful tool for understanding mathematical concepts but we should not try and make the manipulative fit the concept. We should be thinking about the concept that we are teaching first and then choosing manipulatives, where appropriate, to support the teaching of a concept.

(3) **Jo Morgan**, 2019. *A Compendium of Mathematical Methods* Woodbridge: John Catt.

We should get to know the strengths and the limitations of methods that we use. Alternative methods will have their own strengths and limitations and we should know which method may work best for which mathematical situation.

(4) **Jemma Sherwood**, 2018. *How to Enhance Your Mathematics Subject Knowledge: Number and Algebra for Secondary Teachers (Oxford Teaching Guides)*, OUP.

Using correct mathematical language throughout the learning process will help students to build an understanding. We should be confident to use 'complex' language earlier on in the introduction of a mathematical concept.

(5) **Dan Pearcy**, 2020. *Mathematical Beauty: What is Mathematical Beauty and Can Anyone Experience It?*, Woodbridge: John Catt.

Exploring meaning in mathematics is a big motivator for students. By exploring meaning, we can increase students' mathematical curiosity.

(6) **Michael Pershan**, 2021. *Teaching Math with Examples*, John Catt.

Narrating a process whilst modelling an example is a powerful way of scaffolding, enabling students to make links between pieces of mathematical knowledge. Narrating metacognitive behaviours can support students to become better problem solvers.

(7) **Anne Watson**, 2021. *Care in Mathematics Education: Alternative Educational Spaces and Practice*s, London: Palgrave Macmillan.

To care for our students involves giving them 100% of our attention. To care well, we must actively listen to their responses and support accordingly.

(8) **Ed Southall**, 2021. *Yes, But Why? Teaching for Understanding in Mathematics*, California: Corwin.

Students' ability to communicate mathematical ideas can be built by using consistent language. Being mindful when using mathematical terminology helps students understand mathematical vocabulary. This then enables them to use key language in the communication of their own ideas.

(9) **Craig Barton**, 2020. *Reflect, Expect, Check, Explain: sequences and behaviour to enable mathematical thinking in the classroom*, Woodbridge: John Catt.

Periods of silence in the classroom reduce cognitive load and offer students the opportunity to focus their attention on the intended task. If a student is sitting silently, not writing, it does not mean that they are not thinking. At the right times, silence is golden for mathematical development.

(10) **Chris McGrane & Mark McCourt**, 2020. *Mathematical Tasks. The Bridge Between Teaching and Learning*, Woodbridge: John Catt.

It can take time to learn which mathematical tasks work, why they work, and when they work. There are lots of tasks available; we should use books like *Mathematical Tasks* and *If I Could Tell You One Thing* for sourcing appropriate tasks that enhance the learning of students for what we want them to learn.

Books can offer the light when we find ourselves in darkness

There are many more books out there and this list is not intending to be definitive, but simply what we have looked at so far. The key takeaways above are our own; they are from a combination of reading the section, the rich discussions with the authors, and further reflections afterwards.

We are professionals, making a plethora of micro-decisions every day. It is a social job where we need to know our students, adapt to needs, and show care. Evidence from research does not offer us an instruction manual for how to navigate this incredibly complex job. The books cannot tell us how to teach. But, as Bradley Busch put it so wonderfully in *The Teaching Life*, "On our own, we are all separately bumbling around, lost in the dark to our own experiences. With research, we at least have a lamp and a map to help guide us" (Jones & Macpherson, 2021).

Reference

Jones, K. & MacPherson, R. 2021. *The Teaching Life. Professional Learning and Career Progression.* Woodbridge: John Catt.

7   *Teaching Ratio and Proportion*

Dawn Denyer

As a mathematics teacher of some twenty-seven years, I have taught the English secondary school curriculum many times over and in many "varieties". I started my mathematics teaching within the confines of the 1991 National Curriculum, then introduced the Key Stage 3 National Strategy as a deputy subject leader. As a subject leader it became my responsibility to support others to develop and enhance their teaching. Later I was involved in writing materials to enhance what was available, I undertook my Masters in Mathematics Education and engaged an NCETM (National Centre for Excellence in Mathematics) funded Teacher Enquiry Project. More recently I have been involved in my local Maths Hub as a Professional Development Lead and Secondary Mastery Specialist.

Context

As a beginner teacher I taught a group of students who though they struggled with mathematics were intent on a career in nursing. I supported them in class with ways to understand how to work with ratio and proportion as this was the key concept holding them back from being able to work with medicine conversions.

This then encouraged me to look further to ways I could improve my teaching to enable students to make more connections with the topics they were studying. The research that had the most profound effect on me was the work of the CSMS (*Concepts in Secondary Mathematics and Science* research programme) edited by Kath Hart (1981) and the subsequent SESM (*Strategies and Errors in Secondary Mathematics Project*) Hart (1984). The work appealed to me as it offered insights as to how children developed strategies to solve mathematical problems and how these strategies could be categorised, as well as any resultant errors. As I became more experienced, I found myself revisiting the work of Hart, and using the research to help me plan more fully to cater for misconceptions that the students I taught might encounter.

Underpinning all of this has been my interest in the teaching for understanding of ratio and proportion. This interest was first piqued early in my early teaching when I began to realise that, those students who were struggling with ratio and proportion, also struggled with a variety of other concepts, all of which related back to their understanding of unitary ratio and proportional reasoning. They were also hindered by their poor multiplicative reasoning skills, and this is why it became the focus of my Master's dissertation.

The guidance in *The Framework for teaching mathematics: Year 7, 8 and 9* (DfEE, 2001) explained clearly the importance of proportional reasoning: after calculation the application of proportional reasoning is the most important aspect of elementary number. Proportionality underlies key aspects of number, algebra, shape, space and measures, and handling data ... The study of proportion begins in Key Stage 2 but it is in Key Stage 3 where secure foundations need to be established.

The Teacher Enquiry Project I undertook built on the research from my Master's dissertation and echoed the findings of Noelting (1980), Hart (1981, 1984) and Karplus (1983). Students found proportional reasoning problems difficult to solve. They found simple integer ratios easy to deal with, for example, 1:2 and 1:3, as this involved 'simple' doubling or trebling. Ratios in the form 2:1 were also handled reasonably well. Non-integer ratios proved more problematic, as few students could manage to use either a unitary method or a functional algorithm to help them solve problems that used a ratio of 2:3 or 5:3. The difficulty students faced was reflected in the experiences of student Nurses in research by Hoyles (2001) thus taking my own experiences full circle.

I believe that my Master's research demonstrated that learners need to be secure in their understanding of the fundamental mathematical skills before they can advance to solving higher-level proportional reasoning questions. Learners need to have confidence in their ability to use addition and subtraction, as well as multiplication and division. Learners also need to understand fractions as an intensive quantity as well as a quotient, or part of a whole, as discussed by Nunes et al (2005).

My experiences from this study would also lead me to advocate more time being spent discussing mathematics with students; not just going through the motions, but to actually invoke dialogue with students, enabling misconceptions to be unpicked. I believe that encouraging students to discuss not only 'what' method but also 'why' they are using a particular method is particularly important with areas of mathematics that are perceived to be difficult to teach and learn, such as proportional reasoning. Oracy in the classroom enables student to express their understanding. Only then can we truly begin to understand the schemata that our learners have formed and begin to address any misconceptions. This is as true now as it was in 2008!

Applying this in the classroom

Some time has passed since, and the school curriculum support has changed too. The National Strategies have gone, as well as Local Education Authority teacher networks, the NCETM has evolved and local Math Hubs and other subject specialist teacher networks have emerged from Complete Maths #Mathsconf on Twitter; all of which have enabled the promotion and dissemination of techniques to interested teachers.

If I reflect back over this time, the resource that originally changed my teaching, and that of my department was the National Strategy resource "Interacting with mathematics in Key Stage 3: Year 8 multiplicative relationships: mini pack". Not only did this provide a selection of oral and mental starters to help frame the learning, but it also addressed the need to move from additive to multiplicative reasoning. This was the first time that I had seen multiplicative inverse used in this way to model the scaling of numbers, and the use of "awkward" numbers encouraged to help extend the students' understanding of the concept.

Parallel number lines were introduced, as a way to show the multiplicative scale factor of the unitary method. This remains for me today one of the most powerful ways of illustrating what happens when we are looking for a scale factor. In the parallel lines diagram here, we have illustrated the process of moving from 5 to 8. The bottom two arrows demonstrate the unitary method of dividing by 5 and then multiplying by 8. The top arrow represents the previous calculation as a multiplicative scalar of eight fifths.

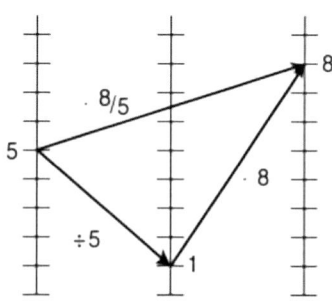

This is a powerful concept and I regularly use parallel lines in this way to help demonstrate the concept of fractions as non-integer scale factors. Anecdotally, I am known in school for my "fractions are our friends" catchphrase. Only recently, I was helping a group of students who were trying to find non-integer scale factors whilst working with similar triangles. Within minutes of showing them how to use this diagram they were confidently able to multiply AND divide by the fractional scale factor as appropriate. Later the same students confidently demonstrated their understanding by finding the algebraic common ratios for the geometric progression $x - 2, 2x - 1, 7x - 8$.

I really do consider this resource to be one of my favourites, as it also introduced two other key teaching aids: the dual number line and ratio tables.

In the early 2000s, counting sticks made their first entrance into secondary mathematics classrooms, although I am not sure how many exist in classrooms today. They were initially suggested as being useful for working with times tables and sequences; however, in the classroom where interactive whiteboards were only just making their first appearance, they provided a handy visual representation of a double number line, ideal for illustrating direct proportion. Again, the Year 8 Multiplicative Relationships pack documented how to use the counting stick to simultaneously generate two sets of multiples to represent direct proportion and equal sets of ratios. (Dual numberline department workshop materials are now available from the NCETM website)

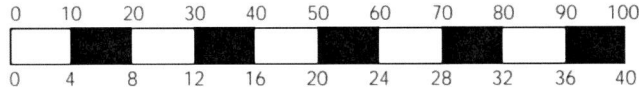

Ratio tables are powerful when understood and used well. They are useful in so many circumstances. Originally I started using them when I was working with obvious situations of direct proportion. The Year 8 resource pack introduces the tables here with the question: "What numbers could go in the boxes? Is there a unique set?". Students quickly grasp the idea if they place a number in the top left they can generate the three missing numbers. The next task is to challenge them to place a number in one of the other positions and work backwards.

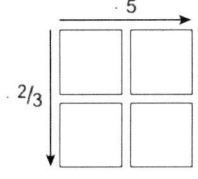

The supplementary notes accompanying the resource pack provides an insight into the many places this technique can be useful and illustrates examples from the National Curriculum tests involving proportion, ratios and pie charts.

The Standards Unit: Improving Learning in Mathematics resources were produced in 2005 as a response to the Smith report: *Making Mathematics Count* (2004). The materials produced use active learning approaches originally designed for post-16 mathematics but were introduced for use across the secondary phase. These materials are well structured and easy to use in the classroom. The unit N6 Developing Proportional Reasoning was designed in recognition that

> "proportional reasoning is notoriously difficult for many learners. Many have difficulty in recognising the multiplicative structures that underlie proportion problems. Instead, they use addition methods, or informal methods using doubling, halving and adding. This session aims to expose and build on this prior learning."

The task encourages the comparision of additive and multiplicative strategies; encouraging the latter, and students are encouraged to then create their own problems, having been introduced to the "box" method of "solving proportional problems in one step".

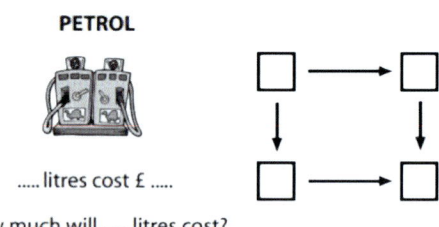

More recently I was involved in an NCETM project on KS3 multiplicative reasoning. One of the tasks we created in our department was a resource

N6 Sheet 1 – *Problems* (page 1)

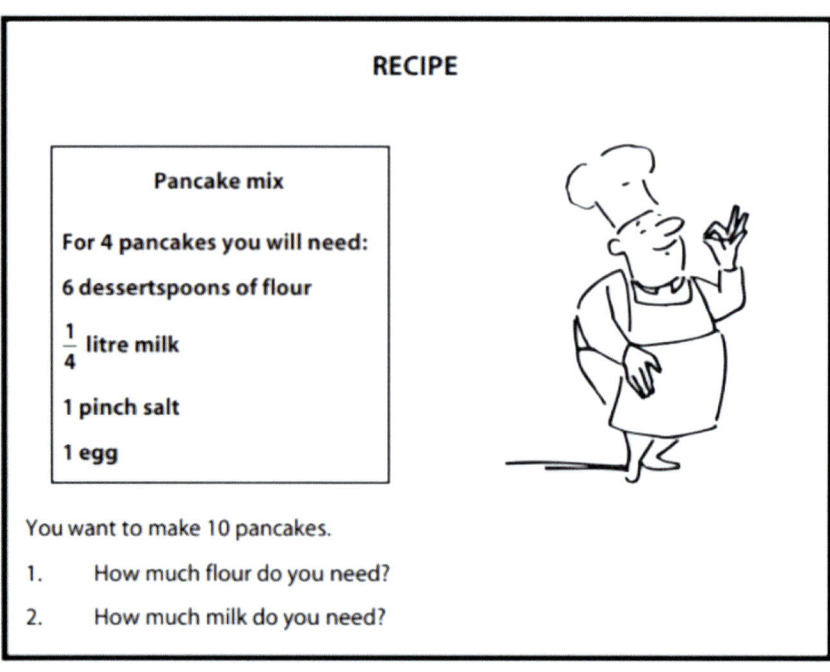

related to this question idea from the Standards Unit. This question was the starter. The task then asked students to work with questions such as "What if I only have 3 dessert spoons of flour, how many pancakes can I make?" Students then created their own situations and questions.

This allowed students to demonstrate their understanding and they created some complex questions to challenge their peers! If we ever had any fear about using 'awkward' numbers, this certainly illustrated that their ability to deal with them relied on their understanding of the processes involved. In these scenarios the situations students face is familiar to them, this allows the mathematical structures to be unpicked more easily. After this task students were more able to answer questions with more abstract settings, including being able to perform complex calculations in a medical context.

3	60
4	

3	9
5	

3	
9	11

2	
7	9.8

The technique of finding within and between strategies using ratio tables was also addressed by the NCETM in one of their Departmental CPD workshops. Worksheets provided questions to discuss as a department to illustrate the concept of ratio tables. They addressed the missing box type questions, as well as worded problems.

A worksheet I have used regularly from this set of materials has a selection of worded questions and relevant ratio tables, thus allowing students to be scaffolded in their selection of the relevant table. This allows students to see how the table can be constructed to help them, before then moving on to create their own table. More recently, I have added another task to my collection of favourite resources and this resource from Don Steward has become another favourite resource of mine. "Boxes" was posted on his site in 2007 having been published initially in 1995.

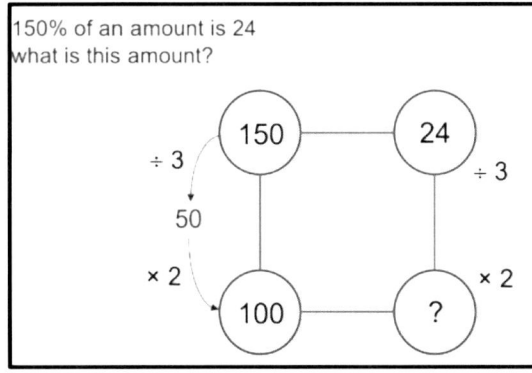

This task involves introducing a common approach to all ratio, percentage and similarity questions, encouraging students to make the connection between the values using multiplicative reasoning.

My involvement with the NCETM introduced me to ICCAMS (*Increasing Confidence and Competence in Algebra and Manipulative Structures*) a four-year research project led by King's College London (from 2008). It investigated ways to raise students' attainment and engagement by using formative assessment. The initial ratio assessment used by the project reminded me very much of my own research project based on the Kath Hart materials. The two lessons with open access look at multiplication in terms of scaling, and model this with the double number line in the context of a map, with scales showing distances in metres and feet; this is then related to the use of ratio tables. The second lesson introduces mapping diagrams and Cartesian graphs, consolidating work from the previous lesson and considering the different ways that multiplication can be illustrated, and yet again demonstrating the importance of making the connections between concepts visible as well as illustrating the mathematics found across the curriculum.

If I recall the content of the mathematics curriculum that I have taught this academic year, the classes I teach have a large impact on the focus; however, the connections between ratio and proportion and multiplicative reasoning are a silver thread running through the taught material. Whether I have been looking at similar triangles and scale factors of enlargement, direct proportion and "Best Buys" GCSE questions, to finding unknown amounts in "reverse" fraction questions with KS3, the same multiplicative reasoning is required. I have taught using ratio boxes and parallel lines, as well as using bar models and dual number lines, each time highlighting the link to the multiplicative reasoning required.

I am still on the journey, along with my students, to find the way that unlocks their understanding. I cannot claim to have found the "perfect" explanation for every student for every situation as yet. However, I hope that having such an arsenal of strategies, I am able to provide an appropriate approach that enables students to engage and succeed on their mathematical journey.

References

NCETM Department Professional Development resource
https://webarchive.nationalarchives.gov.uk/ukgwa/20110505225051/https://www.ncetm.org.uk//resources//10334
ICCAMS, http://iccams-maths.org/multiplicative-reasoning
Dual numberline materials are available from the NCETM website
www.ncetm.org.uk/classroom-resources/secmm-31-understanding-multiplicative-relationships/
DfE *Developing Proportional Representation*
www.stem.org.uk/resources/elibrary/resource/26924/developing-proportional-reasoning-n6
DfEE 2001. *Key Stage 3 National Strategy. Framework for Teaching Mathematics: Year 7, 8 and 9*, DFEE Publications.
DfE 2002. *Year 8 Multiplicative Relationships Mini-pack*, online at www.stem.org.uk/resources/elibrary/resource/29300/year-eight-multiplicative-relationships-mini-pack
Hart, K.,(ed.) 1981. *Children's' Understanding of Mathematics 11-16*, John Murray.
Hart, K., 1984. *Ratio: Children's Strategies and errors*, NFER-Nelson.
Hoyles, C., 2001. 'Proportional Reasoning in Nursing Practice', *Journal for Research in Mathematics Education,* 32, 4-27.
Karplus, R., Pulos, S., Stage, E., 1983. 'Proportional reasoning of early adolescents', in Lesh, R., Landau, M., (ed) *Acquisition of Mathematics Concepts and Processes*, Academic Press, New York.
Noelting, G. 1980. 'The development of proportional reasoning and the ratio concept. Part I – Differentiation of stages', *Education Studies in Mathematics*, 11, 217-253.
Nunes, T., Bryant, P., Pretzlik, U., 2005. 'Children's insights and strategies in solving fraction problems'.
Steward, Don, 'Boxes', online at
https://donsteward.blogspot.com/search?q=boxes

8 Choosing the Right Tasks for Your A level Mathematics Lessons

Susan Whitehouse

The designer William Morris once famously said, "Have nothing in your houses that you do not know to be useful or believe to be beautiful." A corresponding principle for maths teachers might be to have nothing in our lessons that we do not know to be useful or believe to be important. But how do we ensure that the tasks we choose for our A level maths lessons stick to that principle?

Here I will use the words "task" and "activity" to mean anything that the students are spending their time on during the lesson. This might be a matching activity or collaboration when solving a problem, but equally well it might be completing a textbook exercise or focusing on the teacher modelling an example. And I will use the words "aims" and "objectives" to mean whatever the teacher is trying to achieve within a lesson or over a series of lessons, with no real distinction between the two words!

A while ago, sharing learning objectives for individual lessons was very big in education. It was compulsory in most institutions for teachers to explicitly share their lesson objectives with their students, even A level students. Lessons began with teachers displaying precisely what they expected them to have learned by the end of the period. But then came the pushback. Maths teachers pointed out that it is often not appropriate to reveal to students at the beginning of the lesson the mathematics that you hope they will discover during the course of that lesson … that can be like telling the punchline of a joke first. We also argued that to have specified our lesson objectives to the students puts pressure on the teacher to deliver these objectives regardless of the direction the lesson takes, and thus it can be a barrier to responsive teaching. We pointed out that, by their very nature, lesson objectives tend to focus on short-term acquisition of skills, and this can lead teachers to neglect the longer-term development of their students as mathematicians.

Now it is much less common for schools and colleges to insist that teachers share their learning objectives with students. However, it is still clearly a vitally important part of planning any lesson that the teacher does identify for themselves what it is that they are trying to achieve when they deliver that lesson. This might be something that can be achieved within this one lesson and it might be something to be worked towards over a series of lessons, or even over the entire course. Most likely it will be a mixture. But without a clear identification of what you want your students to get out of the lesson that you are planning, you are not going to be able to choose appropriate tasks.

The A level maths specifications consist of not only detailed subject content but also the overarching themes of proof, problem-solving, argument, language and modelling. Likewise, the aims that the teacher has for any A level maths lesson will probably be a mixture of specific mathematical content objectives and wider aims for their students' mathematical development. But for any given mathematical topic, these aims will not only be different for two different teachers, but also different for the same teacher with two different groups of students. They will depend both on the needs of the learners and also on the vision that the teacher has for the development of them as mathematicians.

Let us think about the specific objectives regarding mathematical subject content first. Suppose, for example, that you are delivering a first lesson on the equation of a circle. Depending on your pedagogical beliefs and upon how you construe the interconnection of the different mathematical ideas, your primary aim for this lesson might be for the students to achieve algebraic fluency relating the equation of the circle to the radius and the coordinates of the centre. But it might be for the students to achieve graphical fluency relating the equation of the circle to the appearance of the graph. It might be for the students to discover for themselves the standard form of the equation of a circle using graph transformations, or using Pythagoras' Theorem. You might aim for more than one or even for all of them. Or you might have an entirely different aim. I am not going to debate here which of them are the correct objectives for your first A level lesson on the equation of a circle; that is an argument for another time! But I will say that I do not believe that you can decide on the best structure and activities for your lesson unless you are clear on what it is that you are trying to achieve.

Now let us think about the wider mathematical aims for your lesson. The exam board specifications state that the overarching themes should be delivered across the whole of the detailed content. However, too often this can mean that the development of these skills is not properly addressed at any point in the course. As well as the specific mathematical techniques and concepts, your aims need to encompass developing your students as mathematical problem-solvers who can model, who can discuss their mathematical thinking, who can conjecture, and who can form coherent and logical mathematical arguments and proofs both verbally and in writing. Teachers need to be clear on which of these they want to be their focus within a lesson or series of lessons, and that will very much depend upon the strengths and weaknesses of each different group of students at any given time.

I am often asked to recommend an activity or task for a particular topic of the A level maths course, and I never feel able just to give an answer without more information. I always need to know more about the teacher's vision of the lesson before I can suggest an appropriate resource. So, does this mean that we can never identify activities as good resources, and add them to our schemes of work? Well, certainly I think that we can identify tasks that are not good! While we can tolerate the odd simple mistake – and indeed we can use such errors in our teaching – we should steer well clear of resources that are flawed, unintentionally confusing or that reinforce misconceptions.

But it is harder to identify absolutely which tasks are good, because different "good" tasks achieve different things, even for the same topic. A task that is designed to promote graphical thinking and collaboration is unlikely to be the one to use if your priority for a lesson is for your students to develop individual fluency in the algebraic skills needed for a particular topic, and vice versa. When adding resources to a scheme of work it is helpful to identify the pedagogical focus of each of the tasks and, if possible, to have a choice of activity for each topic to allow teachers to choose the one that best matches their aims. As a department, you might decide together when it is appropriate to focus on particular aspects of the students' mathematical development, within the scheme of work. But this should always come with some degree of flexibility, so that individual teachers can meet the specific needs of their learners.

If we are clear about our aims for the lesson when choosing the tasks, does this mean that our lessons will always contain only things that are mathematically useful and/or important? Of course not! Even if we have a very clear vision for the lesson, we can never be entirely sure of how it will play out on the day, in the classroom, with our group of learners. But by having identified exactly what it is that we are trying to achieve, we can select activities that direct the students' attention to what we want them to be thinking about and that encourage them to work in the ways we think are most important. And it means that we are more likely to be able to get the lesson back on track if we feel it is going in an unhelpful direction. There is no guarantee of success, but it gives us the best possible chance!

9 Invest Time and Thought into Your Demonstrated Examples

Nikki Rohlfing

Scenario: Imagine you've been asked to teach adding/subtracting algebraic fractions. What four examples would you choose to demonstrate this topic? Give yourself a minute to contemplate or write them down before continuing to read. We will return to them at the end!

When I first started out in teaching, I was excited to take the class on mathematical explorations, desperate to show them the beauty and joy of mathematics. In hindsight, though, the thing is that I had not prepared them too well. I had no doubt rushed through poorly chosen examples and assumed understanding and pupils' ability to make 'simple' connections. This lead to troublesome mathematical expeditions, all-round confusion, and loss of pupils' confidence in maths. Only recently, David Spiegelhalter, Hannah Fry and Matt Parker all spoke on the radio about how confidence made a significant difference in learning maths.

This is why, not far into my teaching, I felt I had to pay more attention and give time to those fundamental building blocks and connections; to develop a stronger platform from which we could then successfully go 'off-piste' in lessons. Around this time I also attended a seminar on Shanghai mastery which helped influence these decisions. Specifically, I changed how I think about the examples I use to introduce a new concept and how much time I spend going through them with a class.

Let us return to adding/subtracting algebraic fractions to help explain how I might approach this. (Note that I don't expect you to do it exactly the same, and hope it merely serves to help you develop your own personal approach!) I might start with:

$$\frac{x}{2} + \frac{1}{3}.$$

"What makes this different to what you know about fraction addition so far?", I might say, challenging prior knowledge. Similarly, when using harder examples later, "What is it exactly, that makes this harder?" to help pupils pick apart the skills and think carefully about the nuances between examples.

Then, I would talk through all the steps involved carefully and thoroughly. I fear it could sound condescending to go into detail here, and perhaps my previous sentence already seems 'obvious' as well but do invest some time in thinking about the precise terminology and structured layout of your examples. The other aspect I would always do with my examples is to physically work through them on the board myself, rather than click through a PowerPoint. I really think it's valuable for pupils to have it clearly demonstrated, from the structure of how they should lay it out to the terminology used, and supported by a running commentary on your insight into what you are thinking at each step.

Once complete, it's time to pause. Allow the information to settle and see if anyone might have questions. After this, I would pick on a pupil to explain once again the steps involved, on the same question, and require that they also use the correct terminology, and explain in detail. I might add questions to help probe for understanding: "Why are we multiplying the numerator and denominator of the first fraction by 3? Why is that allowed?" Really build on this fundamental platform before moving on. Perhaps even ask another pupil or two to also repeat the steps. This ensures that when you are demonstrating an example, they know to pay attention.

This idea of creating a narrative with your explanations, with small steps shown and explained in detail, as well as time for thinking/questions has been shown in studies to be a crucial part of effective explanations in maths education.

I make a big deal out of these initial examples where a new concept is involved. It may seem like a waste of time, and the fear of not making 'rapid and sustained progress' (to quote some outdated ideas) might still be prevalent in some schools. I do find though, that demanding pupils' utmost attention at this key stage will ultimately create more fluidity, success and fun in maths. If it helps, I actually only use a blue board pen when doing these important examples, to further highlight their difference and significance compared to regular board work "Okay, I've got the blue pen, so let's pay attention as we delve into a new topic". Again, I am horribly aware of how dull this can be, but the value of understanding and confidence the pupils gain will be well worth the investment!

After the first example, it's time to do a second: "Right, who's ready for level 2?" Given how long you will spend on these examples, it really is important that they are thought through carefully to avoid wasting time accidentally, for example, by doing the same question again but merely with different numbers. Which of these questions would make a good next example in a natural progression?

$$\frac{x}{2}-\frac{1}{3}, \quad \frac{x}{2}+\frac{x}{3}, \quad \frac{2}{x}+\frac{1}{3}, \quad \frac{x+1}{2}+\frac{1}{3}, \quad \frac{y}{2}+\frac{1}{3}, \quad \frac{5x}{2}+\frac{1}{3}.$$

The key thing for me is that a next example shows something different, something that pupils are unlikely to be able to infer themselves. Looking for differences and similarities between consecutive examples is also really beneficial for understanding and helps pupils decode scenarios. My next example would also depend on the ability of the class, and how much of a jump they might (not) be able to do. In that sense, there is no 'right' answer, but to offer my thoughts, I might actually choose something like:

$$\frac{5x+1}{2}+\frac{1}{3}.$$

I might start by picking a pupil and asking, "Why is this one harder than the first?" Analysing this helps develop understanding of the complexity of layers involved, and it's amazing how pupils are then able to approach horrific looking questions because they have the ability and understanding to break them down and tackle them with confidence. For example,

$$\frac{\frac{x}{2}+\frac{3x}{x-1}}{\frac{4}{x}-\frac{2x-1}{2}}$$

is clearly not one to use as an example, but a healthy challenge for pupils who have had time to thoroughly engage with everything that has happened during the explanatory set of questions.

I'll admit that this approach does lend itself to some topics better than others, especially algebra. However, I do use it with all topics and all classes, and believe that this has helped my pupils more than how I used to do things before. Perfecting those chosen example questions is a never-ending challenge which I see as an enjoyable puzzle part of teaching, as each class has its own subtly-different requirements and prior knowledge.

Finally, let us now return to the scenario right at the start and the four examples you chose. Having read this chapter, would you want to go back and change any of them? If yes, then hopefully this chapter has been helpful; and if no, then please send me your chosen examples – they sound perfect!

10   *Teaching Problem Solving*

Sheena Flowers

I look back to my early career and I remember at the end of a topic, I would give students some "problem solving" questions based on what they'd just learned. I thought I was teaching students problem solving skills but I really wasn't. They didn't have to identify the bit of maths they might need to solve the problem as it was the topic they'd just been taught. I'd have spent a few lessons teaching Pythagoras' Theorem then bung some wordy "problem solving" questions on the board about ships sailing around or something similar and think that was it – job done! We can move onto Trigonometry. There is nothing wrong in giving students these types of questions as part of a topic but we are kidding ourselves if we think this is teaching problem solving skills.

A number of years ago I attended a session by the AMSP on preparing students for University entrance exams. The main thing that stuck from that session was that if you are teaching the skill of solving problems in maths, the maths content needs to be familiar to the student. They shouldn't have to apply maths they are not confident with or this will be the focus, not the problem solving itself. This was like a lightbulb moment and completely changed the way I approached problem solving in maths. The sweet spot seemed to be using content students had learned two years previously and certainly not something you had recapped recently. That meant when teaching these skills to year 7 students I had to look at the topics from the Primary Curriculum in year 5. Even then I had to check the students were secure in the skills they needed to solve the problems I was giving them. This meant I was explicitly working on their problem solving skills rather than the topic.

Another thing I took from this session was to let the students be stuck. So often I have found myself intervening far too early when students are struggling with problems. I would give them hints on the maths they need, not asking the right questions of them to guide them towards a conclusion. Key questions I should have asked were things like, "What information do you have?" and "What can you work out?" The questions had to get students to think about what they could do with the information they'd got, even if it led them to a dead end. Most importantly, allow them to head down that dead end if need be.

A great place to start on problem solving skills with a class is to use Goal Free problems. Give the students some information and ask them to find out as much as they can. For example, using the diagram below, you can give students values for A and B and tell them to explore. There are then further opportunities for generalisation. What would change if A was increased/decreased? What about B? What if they were negative?

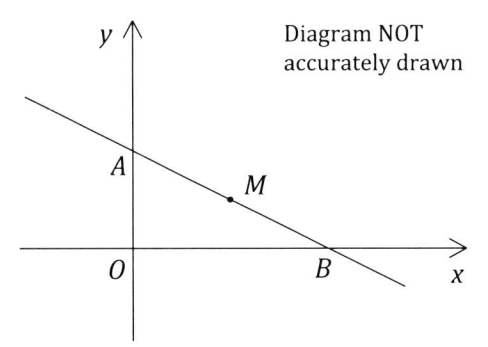

Diagram NOT accurately drawn

It could be as simple as giving two lists of numbers with headings like "year 7" and "year 8" or "class 7G" and "class 7H" and asking students what they can work out from the data. This way there is no right or wrong answer. There are no time constraints and it doesn't matter if one student comes up with completely different things to another. You can then get the students to come up with questions they could ask and what the response would be. To make it clear, this is not about inquiry-based learning. The students will already be familiar with the maths needed to come up with some conclusions and questions. They are learning about what tools they have in their mathematical toolkit and how they might apply them to the information in front of them. Using goal free problems is low stakes for students, there isn't a right or wrong answer and gets them focusing on the journey rather than the end result.

On a personal note, I am trying to move away from giving lists of data labelled "boys" and "girls." As the parent of a non-binary child, I know this is increasingly unrepresentative of the young people in front of us. There are plenty of other labels we can use while making the same mathematical point.

Another great source of problems are UKMT past papers from either individual or team challenges. I gave the following problem to some year 12 students. I hadn't attempted the problem myself before the lesson and had launched down a rabbit hole of trigonometry and triangles. One student found a very elegant solution in seemingly no time at all using co-ordinate geometry. We had a great discussion afterwards about the different methods we could have used and the efficiency of each one.

The diagram shows a circle of radius 1 touching sides of a 2×4 rectangle. A diagonal of the rectangle intersects the circle at P and Q, as shown. What is the length of the chord PQ?

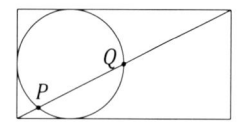

A $\sqrt{5}$ B $\dfrac{4}{\sqrt{5}}$ C $\sqrt{5} - \dfrac{2}{\sqrt{5}}$ D $\dfrac{5\sqrt{5}}{6}$ E 2

I quite enjoy giving students the odd problem I haven't tackled prior to the lesson. I know this will fill some teachers with dread and it certainly isn't for everyone! After letting students begin to find a way in I might start playing around with the problem under the visualiser while they are working. It allows students to make suggestions of things to try. I also think it is good for the students to see me going down dead ends and getting stuck and talk them through my thought processes as I go. Preparing students for STEP is very much like this. I am not going to work through every STEP question, so when a student comes to me for help I don't need to know the answer or the best way to get there. In this situation I am coaching the student to find a way through the problem. Students often go through school without witnessing their teacher struggle with a problem. It certainly isn't something we tend to "actively" model but I think there is a lot of value in showing students how we deal with being stuck.

More recently I have had students preparing for the BMAT exams and finding the time constraints and lack of calculator quite a barrier to their performance. Whilst in a classroom we want to focus on the journey and eventually reaching a conclusion, the fact of the matter is that students are going to be judged on what they can do under pressure and in a limited amount of time. Taking the same problem and trying to find as many different ways to the solution as possible is another approach that can get students tackling problems using different mathematical tools. They will then get better at judging the efficiency of different methods and therefore choosing the best one in the circumstances. The issue my students were having with the BMAT problems wasn't that they didn't know how to answer the questions, it was selecting the most efficient method and selecting it quickly. Another issue we found was the lack of speedy arithmetic skills. Always having a calculator at A level meant they were very much out of practice using written calculation methods.

Problem solving is all about students building up a toolkit of things to try. If you are preparing students for university entrance exams there are some common themes and techniques like spotting the difference of two squares and writing a number with digits xy as $10x + y$. Without exposure to different types of problems, students would find some of the STEP/MAT/TMUA problems very difficult to access. Cambridge University have produced a STEP preparation pack for students which includes some really fun problems that also guides students to learning some useful techniques for tackling maths problems. Each assignment starts with a warmup, goes into a preparation task and then a STEP question. There is then a warm down at the end of each assignment, often containing a maths problem with surprising results. If we do not expose students to lots of different types of problems and techniques this becomes a barrier to them progressing.

This warm down problem from the STEP preparation pack is a great problem with a surprising result because it does not rely on the radius of circle C (providing it is greater than a). It is also accessible to A level students (and even GCSE students).

> The diagram shows a circle C with centre O, and a rod AB the ends of which can slide round the circle C (so that AB is a chord of C). The radius of the circle is R and the length of the rod is $2a$.
>
> As the rod slides around C the point P, which is a fixed distance b from the centre of the rod, traces out a circle with centre O of radius r.

> Show that the area between the two circles is $\pi(a^2 - b^2)$.

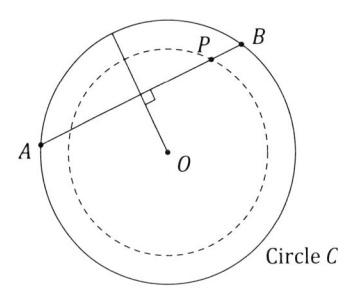

An important point to make here is that problem solving is for all students regardless of year group or ability level. Problem solving is a life skill. The Junior Maths Challenge has some lovely questions to prompt class discussions and get students thinking about how to tackle them. This problem for example, can be scaffolded with some squared paper. You could ask about areas and fractions as well and extend it. For example, what if you add another square length 6?

The diagram shows five squares whose side-lengths, in cm, are 1, 2, 3, 4 and 5. What percentage of the area of the outer square is shaded?

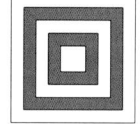

A 25%   B 30%   C 36%   D 40%   E 42%

Practicalities often mean we are inevitably also preparing students for exams. Using past paper questions is also a great way to get students thinking about what the question is asking, what maths they need to apply. One question that took me far longer than it should have was near the beginning of a recent Edexcel foundation paper:

> Draw a quadrilateral with no lines of symmetry and rotational symmetry of order 2.

Looking at some of the papers we requested a Review of Marking, quite a few students also struggled with this question and many left it blank. This was very early in the paper and what struck me was that students would have been awarded a mark for either drawing a quadrilateral with no lines or symmetry or drawing a quadrilateral with rotational symmetry of order 2. This is accessible to most students and had the question been "Draw a quadrilateral with rotational symmetry of order 2", I feel most students would have managed to draw a rectangle. Another aspect of teaching problem solving is the credit you get for trying things along the way. Specifically in an exam context, the answer is worth 1 mark – it is what you do to get there that gets you the most credit.

I really enjoy solving problems. Instilling this in the students we teach is not an easy task. Showing them it is ok to struggle and ok not to get to the answer often goes against what they believe maths to be, a subject where things are either right or wrong. Problem solving is also about resilience and perseverance. Students get a confidence boost when they are successful but will often give up easily, and changing this mindset is a battle worth having. In the classroom, where we have so much content to get through in so little time, it can be difficult to find the time to let students play around with maths. However, the benefits of teaching problem solving as a stand-alone skill will be seen throughout their maths lessons and beyond.

I'll leave you with a couple of problems from my two favourite problem setters, the amazing Catriona Agg and Ed Southall, both brilliant sources of great geometry problems that I share with my students. I have spent many enjoyable hours racing against my partner to solve their problems, and love sharing them with my students too.

Each square has area 1. What's the area of the large rectangle?

What's the area of the semi circle?

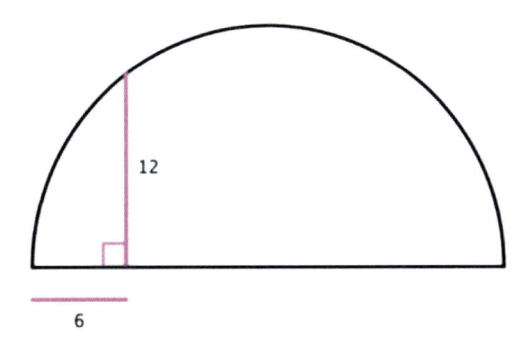

12

6

References

Catriona Agg [Twitter] @cshearer41 and Ed Southall [Twitter] @edsouthall
Goal free problems https://goalfreeproblems.blogspot.com/
Cambridge University Step Preparation https://maths.org/step/
UKMT, 2017. Senior Maths Challenge Paper
www.ukmt.org.uk/sites/default/files/ukmt/senior-mathematical-challenge/SMC-2017-paper.pdf
UKMT, 2022. Junior Maths Challenge Paper
www.ukmt.org.uk/sites/default/files/ukmt/JMC_2022_Paper.pdf

11 *Planning and Teaching for Coherence*

Kathryn Darwin

When I first started teaching, I lived from lesson to lesson for each class. I knew what was coming up in the scheme of learning and the number of hours I had to teach it, so I obeyed it. I'd plan a 5-hour long block on solving linear equations, then three more hours on "nth term", and so on. Each one had a definite start and end-point, though that often meant shuffling slides around on a daily basis once I'd 'taught' each hour long block – not once thinking this was a very difficult way to be doing things. But my students were able to do what I was asking them to do, so I thought myself a pretty good teacher.

Alongside my self-imposed workload, there were other issues; without the context of the lesson, students couldn't 'perform' anymore. My students' success was short-term and focussed on only a small part of the curriculum. My end-of-term assessments showed lack of recall and misconceptions at every turn. They showed students *sometimes* remembering what to do, but other times leaving blanks, even where they had successfully met similar mathematical ideas elsewhere. Students multiplied by two instead of squaring. When factorising quadratics, they 'mixed up' the coefficients needed to find the correct sum and product. They could split £20 between Anne and Bob in the ratio of 2:3, but couldn't tell me what Bob got if Anne got a certain amount.

I came to the realisation that, I believe, most maths teachers come to at some point in their careers – I had always been good at and enjoyed maths, so I made the connections myself. At school I wanted to and was confident in making links and generalisations, even if I hadn't been explicitly taught them. I'm sure we've all had a class of mini-mes who can do this (mine were 8X1 – they loved maths, just like me, and knew they could derail a lesson by asking about types of infinity on a weekly basis). But not all students will have this natural affinity for maths. Many of them don't enjoy it as they see maths as a set of totally unrelated topics, with multiple steps to remember to solve each problem within them. I would argue that even some able students see it like this too – they're just good at the steps!

At that time, I was providing my classes with nothing more than these unconnected ideas and expecting them to fill the gaps for themselves. I'd assumed this would work for them, because it had for me ... it was unfair and unrealistic. They could not possibly see "The Big Picture" if I wasn't showing it to them. How could I make sure they could confidently calculate and use square numbers, remember how the area model could support their factorisation of quadratics, or see that moving the £20 from the total to Anne just meant a change of perspective? And that all of these were fundamentally multiplication problems? Something had to change if I wanted all my students to have access to the 'map of maths' I had in my head.

I had to alter how I planned to reflect this. I knew my use of each hour-long lesson as a discrete unit was contributing to this. I needed to move away from treating each hour with my classes as something new and unconnected, or rushing to get through all I had planned for the lesson. So I started creating 'mega-lessons' for each topic in the scheme of learning. Taking a whole topic and considering *everything* I could incorporate; from pre-requisites to boundary examples, the order of my examples and tasks to the variation I used within them. At La Salle's MathsConf14, Dani Quinn said, "Don't teach now something you will contradict later" and it has stuck with me ever since – I started planning for the development of a concept and making the links across maths more explicit to my students. These mega-lessons then formed the basis of my teaching, picking up where we left off each lesson and building up concepts and their connections gradually, with common language and representations.

I did not know it at the time, but I had begun planning for coherence. In my work as a secondary Mastery Specialist with the NCETM, I came across the 5 Big Ideas, where coherence is described within mathematics lessons as "small, connected steps that gradually unfold the concept, providing access for all children and leading to a generalisation of the concept and the ability to apply the concept to a range of contexts." I see this definition as a 'how to' in two parts, micro- and macro-coherence. Macro-coherence allows students to see "The Big Picture" through use of a carefully-structured curriculum, which exposes the links between number, algebra, and geometry, including representations and structures which can be revisited and reused as ideas increase in complexity. In practical terms, this means carefully crafting lessons with well-thought-out examples, variation and tasks which allows students to build micro-coherence within a single concept or topic, before connecting it to the wider subject.

The Collins dictionary definition of coherence is "a state or situation in which all the parts or ideas fit together well so that they form a united whole." I much prefer this one – it moves away from the mechanics of 'how' and back to the 'why'. This is what I have always wanted for my students and why I began to change my practice years before. I wanted them to have a deep and connected knowledge of mathematics and to be able to use this to work out what to do, instead of remembering, or misremembering, some disjoint steps.

This can seem idealistic and difficult in practice, so I will now illustrate these ideas, with an outline of my approach to teaching multiplicative reasoning. Watson et al (2013) describe it as a 'medium term project' in curriculum and lesson planning, as it appears explicitly and implicitly throughout the Key Stage 3 and 4 curricula. For my students, this begins in Year 7, where all students are exposed to the concept that any two numbers can be connected by a single multiplier. It is important to note that at this stage, they have already studied factors and multiples, and some fractions – most notably considering multiplication by fractions as fractions of an amount and then division as multiplication by the multiplicative inverse or reciprocal. The scheme of learning was deliberately designed in this way to allow these ideas to be continually reinforced and linked to proportionality.

We begin with a familiar context; money! In answering problems such as "If 5 apples cost £1, how much do 7 cost?" many students instinctively begin working with a 'unitary' method. In what Lamon (1993) considers 'pre-proportional reasoning', students find the cost of one unit and then 'multiply up', mirroring ideas of 'for each/every' introduced in Key Stage 2. Throughout, problems are modelled using bar models, double number lines and ratio tables, as shown in below.

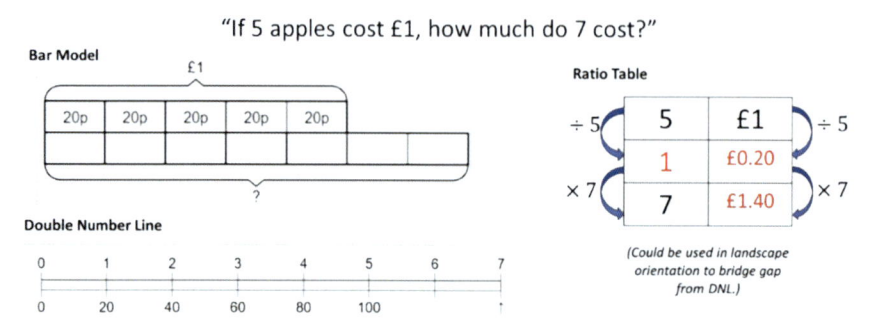

"If 5 apples cost £1, how much do 7 cost?"

Examples then build in complexity, involving numbers with common multiples; "If 8 apples cost £1, how much do 6 cost?" Though students can still work with the unitary method, the use of a common factor of 2 is much more sophisticated (as the next figure shows). In my experience, this is where they begin to prefer the use of the ratio table, as a result of its efficiency.

"If 8 apples cost £1, how much do 6 cost?"

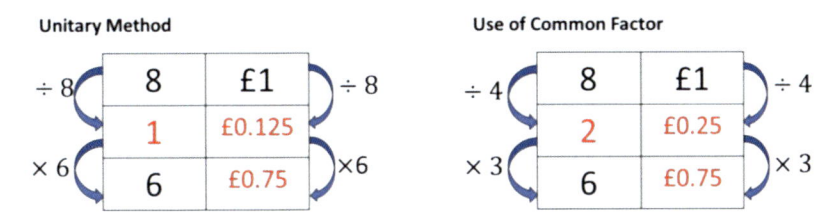

Unitary Method

8	£1
1	£0.125
6	£0.75

÷8, ×6 (left) ÷8, ×6 (right)

Use of Common Factor

8	£1
2	£0.25
6	£0.75

÷4, ×3 (left) ÷4, ×3 (right)

Additionally, this structure makes clear the 'within' and 'between' relationships. 'Between' strategies focus on comparing variables in the same unit, whereas 'within' strategies consider a factor connecting different variables. These are illustrated below. This again increases the sophistication of multiplicative thinking, as per Lamon's framework for proportional reasoning.

"If 10 apples cost £1, how much do 50 cost?"

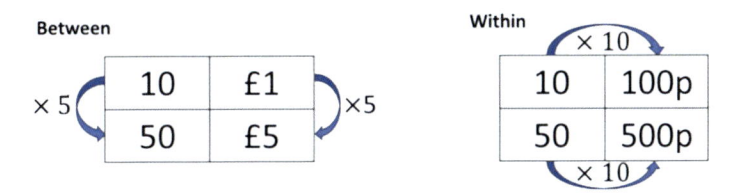

Between

10	£1
50	£5

×5 (left), ×5 (right)

Within

10	100p
50	500p

×10 (top), ×10 (bottom)

Students practise this idea, with context removed to allow for increased fluency, using resources such as 'Boxes' from NCETM (below). These give students an opportunity to decide when 'within' and 'between' strategies are most appropriate. This abstraction also sets the scene for use of ratio tables and multiplicative thinking in various contexts over time.

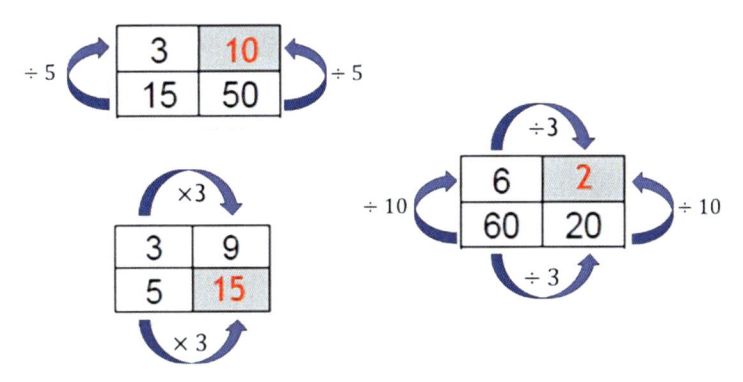

3	10
15	50

÷5 (left), ÷5 (right)

3	9
5	15

×3 (top), ×3 (bottom)

6	2
60	20

÷3 (top), ÷3 (bottom), ÷10 (left), ÷10 (right)

NCETM 'Boxes' resource

To achieve understanding of the original concept, students must let go of "inappropriate whole number thinking" (Norton, 2005). This is possible due to work done on fractional multiplication and division earlier in the scheme of learning, so students are able to combine their multiple steps as one fractional multiplier. Shown in the next figure are both 'within' and 'between' relationships with a single fractional multiplier, and the unitary steps to reach this. Here ÷ 2 is considered as multiplication by the reciprocal, $\frac{1}{2}$, to give the final multipliers.

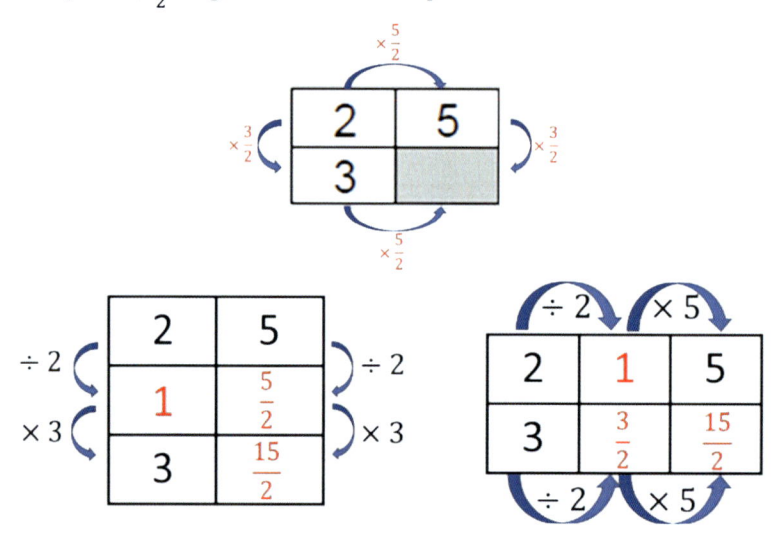

Students are quick to generalise this approach and begin to use '4-box ratio tables' over the unitary or common multiple methods consistently with practice. This becomes embedded within a student's mathematical toolkit and is a consistent representation used throughout the Key Stage 3 and 4 curricula. Thus, the common mathematical structure is exposed within many areas of maths, including:

- Splitting in a ratio
- Fraction of an amount
- Percentages, including percentage change and reverse problems
- Gradient
- Unit conversions
- Compound measures
- Similarity
- Trigonometry
- Pie charts
- Sampling

A selection of these applications are beautifully illustrated in Don Steward's 'Boxes' resources. Though many of them consider numerical calculations only, it is important to note that this approach is not limited to these scenarios. In fact, algebraic proportion problems can also be solved using this representation, ensuring that this concept is not taught as a discrete unit compared to numerical proportion, allowing coherence throughout a student's mathematical journey.

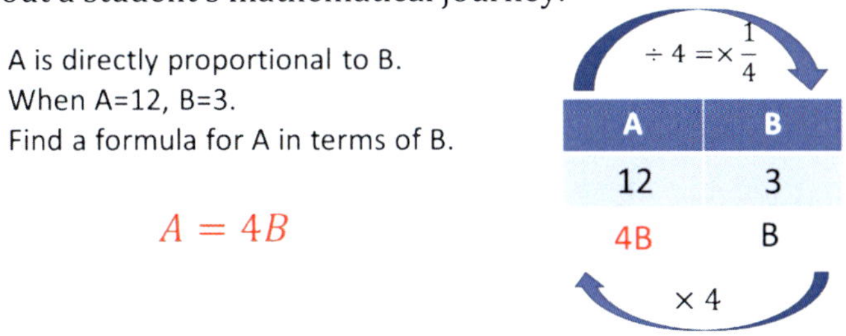

A is directly proportional to B.
When A=12, B=3.
Find a formula for A in terms of B.

$$A = 4B$$

This is an outline of only one idea within the curriculum, and already we can see the complexity of the fundamental concept, but also its connections to other areas of maths. It highlights that mathematics is a magnificent but intricate, interconnected web of ideas that we, as teachers, are confident with – how do we help our students see this beauty and complexity?

My advice is to plan for coherence; always consider what came before, what comes next and how you can best bridge that gap with deep, connected knowledge. So, when you see "negative numbers" in a scheme of work, think about how this appears all the way to "expanding and factorising quadratics" and beyond. Before teaching "simplifying fractions" consider why the unit on "prime factorisation" came ahead of it. If you see "circle theorems" coming up, think hard about which order you present them in and why. Stop planning by the hour and plan a coherent journey through mathematics, and you will guide your students to see the "Big Picture" too.

References

Lamon, S. J. 1993. 'Ratio and proportion: Connecting content and children's thinking', *Journal For Research in Mathematics Education*, 24 (1), 41-61. doi:https://doi.org/10.2307/749385
NCETM 2008. *Departmental Workshops- Proportional Reasoning*, https://webarchive.nationalarchives.gov.uk/ukgwa/20110505161254/https://www.ncetm.org.uk/resources/10334

NCETM, 2017. *Five Big Ideas in Teaching for Mastery*, online at www.ncetm.org.uk/teaching-for-mastery/mastery-explained/five-big-ideas-in-teaching-for-mastery/

Norton, S. J., 2005. 'The construction of proportional reasoning', in Chick, H. & Vincent, J (Eds.), *Proceedings of the 29th Conference of the International Group for the Psychology of Mathematics Education.* 4, 17-24. Melbourne: PME; online at www.emis.de/proceedings/PME29/PME29RRPapers/PME29Vol4Norton.pdf

Steward, D., 2019. *Boxes, Teacher's Notes*, online at https://donsteward.blogspot.com/2012/03/boxes.html

Watson, A., Jones, K., & Pratt, D., 2013. *Key Ideas in Teaching Mathematics; Research-based Guidance for Ages 9-19.* Oxford: OUP.

12 Mathematics is Important for its Own Sake

Mark Dawes

Every mathematics teacher has heard, "When will we ever need this?" from a student at some point in their career. It is tempting to try to justify the teaching and learning of mathematics via its usefulness, either in the real-life experiences of the students, or from elsewhere in the world. In this chapter I will explain why I don't do this, and what I do instead.

Maths is all around us

- Paying for goods online
- The Google search algorithm
- Satnav
- Computer games
- Music
- AI
- The price of aeroplane tickets
- Restocking a supermarket
- Developing new medicines
- Planning and building houses

All of these are likely to be familiar to students, and all of them involve mathematics. Let's look at the first of these in more detail.

When we buy something online, we need a safe way to signal to our bank or credit card provider that we want to transfer some money. Merely sending our details across the internet would be unsafe, so the information is encrypted. Modern cryptography uses very large prime numbers to encrypt and decrypt messages (including credit card data). Aha: here is a real-life, important use for prime numbers! We should store that up for the next time a student asks, "Why do we need to know about prime numbers?" Or should we?

While it is true that this is a use for prime numbers, and that without them we would not be able to buy anything using a card (whether online or in a shop), this isn't actually of interest to students. In the same way that I don't need to understand the mechanics of a car engine to be able to drive a car, none of us need to understand prime numbers to be able to buy things using a debit or credit card. (It is perhaps still interesting to talk about this with students when exploring prime numbers – just not as a justification for studying them.)

The same is true for all of the other uses on my list. Modern music relies on mathematical tools to create and alter sounds, to set up loops and drum-patterns and to distribute songs. But I don't need to understand any of this in order to enjoy listening to a playlist.

In 2003 an American math professor realised that his dog was able to retrieve a tennis ball from a lake by following the route that would get him to the ball most quickly. The dog ran diagonally along the beach (where he could go more quickly) and therefore spent shorter time in the water. This was reported as meaning that the dog "can do calculus" (Dye, 2003). I don't believe we can justify teaching and learning calculus by saying we will use it to help us to carry out real-life tasks like this.

Maths is useful

It is often suggested that we use mathematics when shopping. This rarely happens. Asda reported that some shoppers were asking till staff to stop scanning after their goods reached £30, meaning they were not calculating this themselves (Winchester, 2022). We don't need to add up totals in shops, because the till scans items and automatically calculates the amount we need to pay.

We rarely need to work out 'best-buys' in a shop, because in the vast majority of cases it is the larger pack size that works out cheaper, and anyway the price label on the edge of the shelf states the cost per 100g, or per kg. GCSE exam questions that ask which size packet of corn flakes to buy always focus on price. This ignores other practical issues. The cupboard in my kitchen in which I keep breakfast cereals is slightly too small for the biggest boxes of corn flakes. I deliberately buy the slightly smaller (and marginally more expensive) boxes, because otherwise I need to tear the top off the box to make it fit.

Other practical justifications

There are other difficulties with trying to give practical justifications for the topics we study. If we justify one topic by its practical usefulness, then we probably need to justify every topic in a similar way. There are some areas of the GCSE curriculum that just don't have direct uses.

Some topics might be relevant to certain jobs and roles. Nurses, for example, need a very strong grasp of ratio when they give patients medication. If a student isn't intending to be a nurse then that is not a convincing reason for them.

Mathematics appears in many other subjects. There it is often situated within the subject and can be used in a different way. (For example, when dealing with data in science, the 'range' is given by stating the smallest and the biggest number, which is different from how we use it in maths.)

So why do we teach mathematics?

Mathematics is useful, important and worthwhile, just not in mundane, practical ways. Some of the reasons for teaching it are:

- **Culture**

   There is mathematics embedded in our culture. Percentages needn't exist. We could do everything by using decimals or fractions instead of percentages. Students sometimes wonder why we use 'out of 100' rather than, say, 'out of 1000'. Well the latter does exist. It's 'permille' and uses a symbol that is very similar to the percentage % sign: ‰. In some other countries permille is used. This is reminiscent of the capacity of a can of soft drink. In the UK cans are labelled as containing 330ml, whereas elsewhere in Europe they are shown as 33cl. So there are cultural differences in the way mathematics is used around the world, and we have our own particular ways of doing things.

- **History**

   Mathematics has a long and interesting history. Pascal's Triangle, named for the 17th century French mathematician who used it extensively in his ground-breaking work on probability, was also known in China in the 11th century and India in the 6th century, where it was used as a number pattern and for calculating combinations, shows the different ways mathematics can be applied and developed.

- **Beauty**

    The friezes at the Alhambra palace in Granada, the graph of $y = \sin x$, the roof of the Olympic velodrome: all are mathematical and all are beautiful. There is beauty in some of the mathematical problems we can solve and the ways we use number, algebra and geometry together.

- **Wow-factor**

    Take the reciprocal of the square numbers and then add them up:

    $$\frac{1}{1^2} + \frac{1}{2^2} + \frac{1}{3^2} + \frac{1}{4^2} + \cdots$$

    Continuing this to infinity gives $\frac{\pi^2}{6}$.

    What a surprise! Why should π appear here? Why is it squared? There are so many places in mathematics, in problem solving and in connections between different areas of the subject, where 'wow' is appropriate.

- **Usefulness**

    Notwithstanding what I wrote earlier, there <u>are</u> some occasions when mathematics is useful. We are assailed daily by data and by claims in the media and in politics. It seems important to be able to think critically about the things we read and which politicians, journalists and those with large numbers of social media followers try to use to convince us of a particular viewpoint.

    As part of the sixth form qualification Core Maths, students need to be able to critically analyse claims that are made. My Quibans resources (Questions Inspired By A News Story) include scores of articles taken from reputable news outlets. It is shocking (but perhaps not surprising) how many of them include basic mathematical errors.

    Even here, though, the mathematics used is limited. This sort of critical analysis might involve calculations, using ratio and proportion, calculating with percentages, approximating, interpreting graphs and carrying out statistical calculations. This doesn't come close to justifying the scope of the GCSE mathematics curriculum.

- **Maths is interesting**

So many of the areas of maths we teach at school <u>are</u> interesting. They might involve a mixture of the culture/history/wow-factor ideas I mentioned earlier. Here is an example.

Draw a circle and a diameter of the circle. Then pick a point on the circumference of the circle.

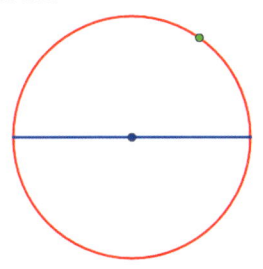

Now join the two ends of the diameter to that point. The triangle formed is always right-angled (unless the point was on one of the ends of the diameter, in which case you have a line and not a triangle).

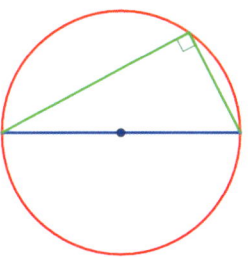

This is known elsewhere in the world as Thales' Theorem, after a Greek philosopher.

Why is the angle formed always a right angle? We could use algebra to explain this, but we can also rotate the triangle about the centre of the circle.

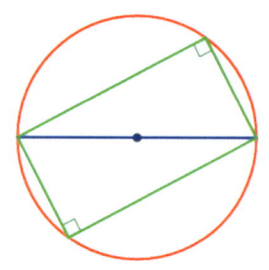

This gives a quadrilateral. The two diagonals of the quadrilateral are the same length (they are both the diameter of the circle) and bisect each other (at the centre of the circle). This means the quadrilateral is a rectangle, and the angles are therefore all 90 degrees.

We study mathematics because it is interesting and beautiful. Any other 'useful' reasons are a bonus!

References

Dawes, M. 'Questions inspired by a news story (QuIBANS)', blog at http://quibans.blogspot.com/
Dye, L., 2003. 'Mathematician's Dog Knows Calculus', abc News (29 May), online at https://abcnews.go.com/Technology/story?id=97628
Winchester, L. 2022. 'Asda shoppers ask cashiers to stop scanning at £30 because of major money fears', *The Mirror* (22 June), online at www.mirror.co.uk/money/asda-shoppers-worried-spending-asking-27300307

13   Noticing and Wondering

Stella Dudzic

When I taught sixth form classes, they often came from different lessons in different places so there would be a couple of minutes at the start of the lesson when they were arriving. They weren't dawdling, it genuinely took some of them several minutes to travel from the previous lesson.

Keen not to waste any lesson time, I started to plan for a short review or introductory activity which could either be displayed or quickly written on the board and which students should start on arrival. Here's one I used when teaching a statistics lesson.

What could the variable be to give a histogram with a shape like the one on the left?

What about the one on the right?

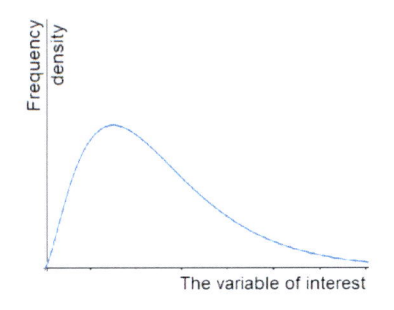

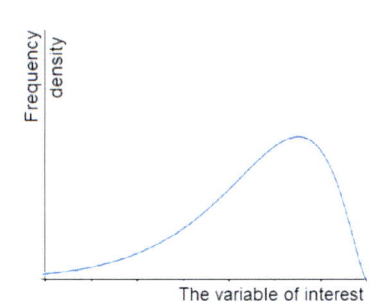

I chose the example above because we had been learning about skewness in histograms and I wanted them to think about what kinds of data might lead to distributions with those shapes. My hope was that, between them, they would be able to think of several examples.

I soon found that my students found it difficult to answer statistics questions where they were asked to make some kind of comment, so introductory activities started to include asking them to make a comment about a statistical chart. I was hoping that they would realise from listening to each other that there was more than one possible answer and that this would lead to them approaching questions asking for comments with more confidence. Students who could not think of any comments were not encouraged by this, so I started to ask them to think of either a comment about the diagram or a question about the diagram, telling them that they might notice something or they might want to ask something and either was fine. I wanted them to understand that mathematics is about asking questions as well as finding answers. This allowed me to get a clearer understanding of what they did, and did not, understand and so plan future learning appropriately.

When I moved into my current job, I continued to use questions like this in teacher professional development sessions. For example, I might ask teachers to think of either a comment or a question about the following chart and then we'd spend some time discussing further.

UK income distribution before housing costs 2010/11

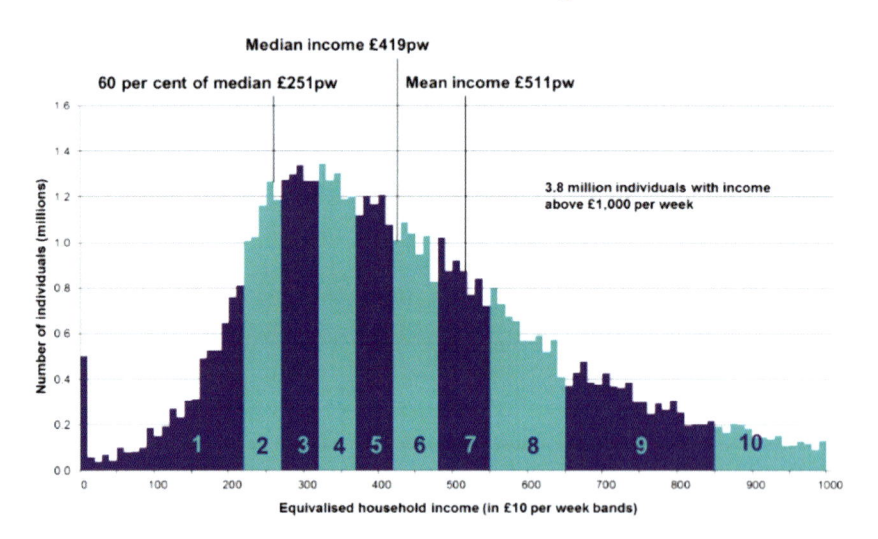

I chose this example because there are many comments that can be made and quite a few questions that can be asked – including some that I don't know the answers to. For example, I still don't understand why there are so many people with no household income. It also links averages and histograms (each decile has the same area).

Using this example allowed me to introduce a type of activity the teachers might use with their students as well as finding out something about their general understanding of averages and statistical diagrams which helped me pitch the rest of the session appropriately.

With the greater focus on problem solving in the current A level and GCSE specifications, I began to read more about problem solving strategies in the classroom; this is when I came across "I notice, I wonder" as a strategy from the US National Council of Teachers of Mathematics (NCTM). It seemed similar to what I had been doing with comments and questions but it was being used as a strategy for problem solving.

Faced with a question like the one below, some students do nothing because they cannot see the complete route to get to the answer. Giving them the confidence to work out something which they can see how to do, and which could be part way to getting the answer is both a good exam strategy (they might get some of the marks instead of none) and good educationally – we want students to do something even if they can't completely solve the problem.

But how to get them to do something, or even to see what they might be able to do? That's where noticing and wondering comes in – either with the diagram alone (so they don't focus on the question but on what they can actually work out) or with the whole question (so they get used to asking themselves what they can do as a precursor to working towards what the question is asking).

> The curve $y = x^2 + 3$ crosses the line $y = 2x + 6$ at points A and B.
> Point C is the turning point of the curve $y = x^2 + 3$.
> Find the area of triangle ABC.

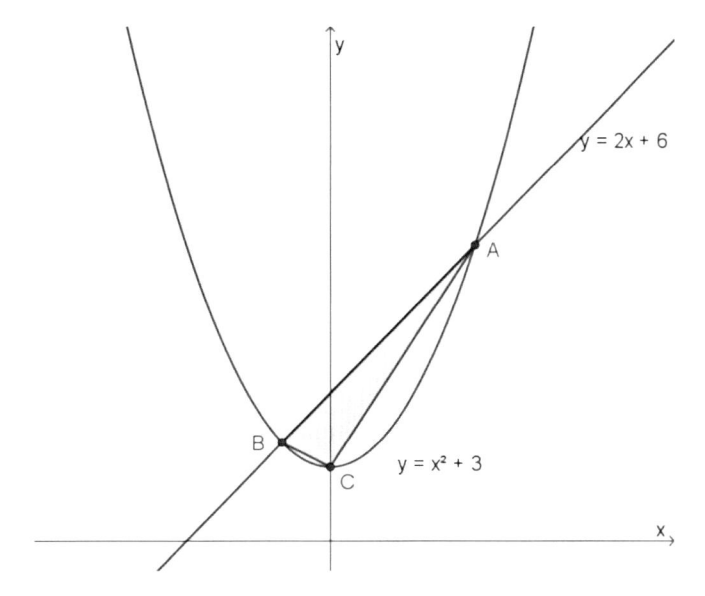

I've chosen this particular example because it's relevant to both GCSE Higher Tier and A level and there are (at least) two stages to the solution:

1. Realising that finding the coordinates of A, B and C will be helpful and going on to find them.
2. Calculating the area of the triangle.

Students may find it difficult to find the area of the triangle even when they have the coordinates of the three vertices – you could ask them which of the two solutions, X or Y below, is better (and why) or even ask them what they notice and wonder about them.

Solution X

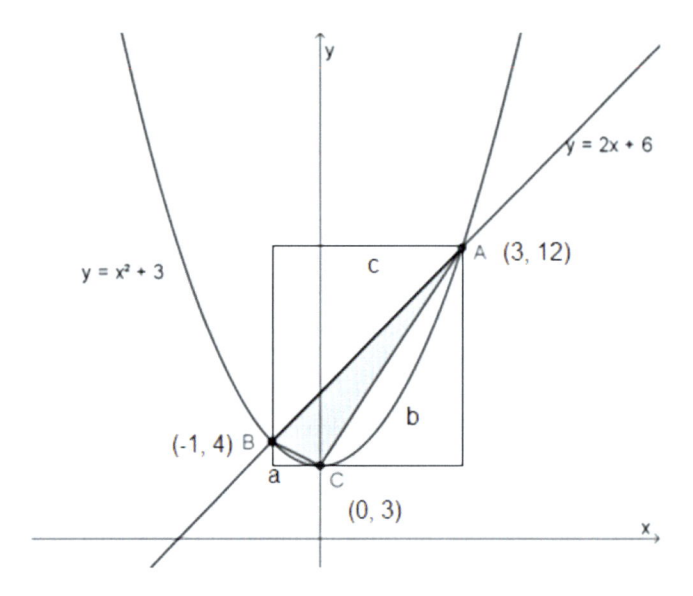

Surround the triangle with a rectangle, sides parallel to the axes and going through all 3 vertices.

Area of rectangle = $4 \times 9 = 36$ square units.

To get the area of triangle ABC, subtract triangles a, b, c.

Area a = $\frac{1}{2} \times 1 \times 1 = \frac{1}{2}$.

Area b = $\frac{1}{2} \times 3 \times 9 = 13\frac{1}{2}$.

Area c = $\frac{1}{2} \times 4 \times 8 = 16$.

Area triangle ABC = $36 - 30 = 6$ square units.

Solution Y

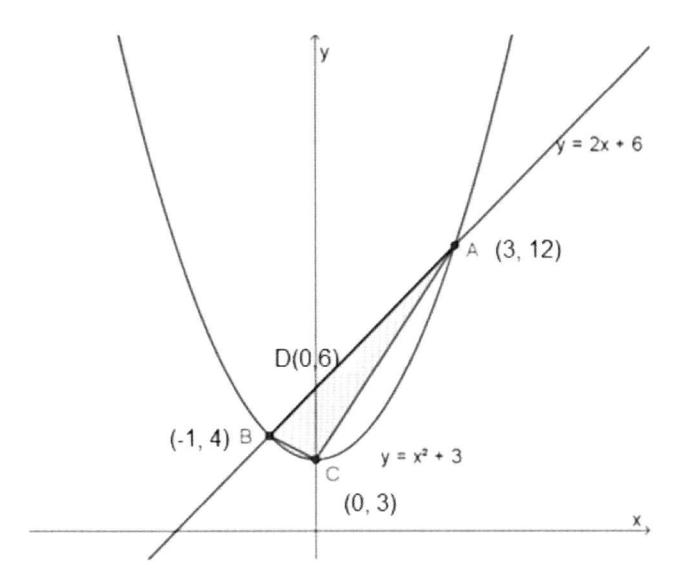

Line $y = 2x + 6$ crosses the y-axis at D (0, 6).
The y-axis splits triangle ABC into 2 triangles.
Area of the one to left of the y-axis is $\frac{1}{2} \times 3 \times 1 = 1\frac{1}{2}$.
Area of the one to right of the y-axis is $\frac{1}{2} \times 3 \times 3 = 4\frac{1}{2}$.
Area triangle ABC = $1\frac{1}{2} + 4\frac{1}{2} = 6$ square units.

Notes

1. Data about the UK income distribution can be obtained from https://assets.publishing.service.gov.uk/government/uploads/system/uploads/attachment_data/file/206778/full_hbai13.pdf
2. NCTM. 'I notice, I wonder', www.nctm.org/Classroom-Resources/Problems-of-the-Week/I-Notice-I-Wonder/

14 *Don't Stop Interweavin'*

Nathan Day

Making connections is fundamental to learning and teaching mathematics. Maths is so full of connections, and it is in the exploration of those connections that much of the beauty and richness of mathematics can be fully appreciated. Despite this, I found that my students often didn't recognise those vital connections and underlying structures, even when they were obvious to me. They saw maths as a collection of many disconnected topics. They saw the surface, but not the depth. They saw the tree's leaves, but not its branches.

One way to reveal these connections is through interweaving. Coined by Will Emeny, 'interweaving' describes the use of tasks that bring together different topics from across mathematics. One simple example of this could be, when studying area and perimeter, having the given lengths be fractions. There are a range of benefits to this. It gives an opportunity to recap operations with fractions in a context where those calculations have meaning. The more involved calculations required to answer these questions encourage learners to be efficient and find shortcuts. And more thinking is required at each stage, eliminating the automaticity that can happen when completing routine questions with easier numbers. I really like using these sorts of questions. However, there is far more to interweaving than chucking fractions and decimals into otherwise ordinary exercises!

Interweaving is all about making connections. One of the reasons that maths can seem disconnected is that key concepts appear in only a few different contexts. An example of this is reciprocals, which learners will see when dividing fractions or when dealing with perpendicular lines. But there are so many other ways you can think about reciprocals and things you can do with reciprocals! Finding the reciprocals of numbers in standard form, for example, requires a deep understanding of both what a reciprocal is and how to perform those calculations with numbers in standard form. It joins up those two dots, and in doing so strengthens them both.

Averages with...

Fractions

Find the mean, median, and range of:

$3\frac{1}{3}$, $6\frac{1}{6}$, and $2\frac{1}{2}$

Area and Perimeter

A rectangle has a width of 3 cm and a height of 2 cm.

Draw a second rectangle so that the two rectangles have a mean area of 13 cm² and have perimeters with a range of 8 cm.

Standard Form

Find the median of the following:

3×10^{-4},
4×10^{-3},
5×10^{-6},
6×10^{-5}.

Surds

John says:

'The mean of $\sqrt{12}$, $\sqrt{27}$, and $\sqrt{48}$ is $\sqrt{29}$.'

Explain and correct the mistake that John has made.

Bounds

Find the upper and lower bounds for the median of the following numbers:

3.5 (one decimal place),

27 (two significant figures),

30 (nearest ten).

Angles

Find the upper bound for the median angle in a quadrilateral.

Is it possible to actually draw a quadrilateral with that median angle?

Find the reciprocal of...

Mixed Numbers and Decimals

a. $2\frac{1}{2}$
b. $13\frac{1}{2} + 21\frac{1}{3}$
c. $0.1\dot{6}$
d. $0.9\dot{8}$

Indices

a. $8^{\frac{2}{3}}$
b. 10^{-4}
c. 0.2^3
d. $\frac{2^3}{2^4 + 2^5}$

Standard Form

a. 1×10^5
b. 5×10^{-4}
c. 2.5×10^3
d. 1.25×10^{-2}

Surds

a. $\sqrt{3}$
b. $\frac{\sqrt{2}}{2}$
c. $\frac{\sqrt{5}+1}{2}$
d. $\sqrt{2} + 1$

Bounds - Find error intervals for the reciprocals of each of the following:
a. 3, rounded to the nearest whole number
b. 2.5, rounded to one decimal place
c. 1.50, rounded to three significant figures

Equations - By forming and solving an equation, find the following:
a. A number that is one quarter of its reciprocal
b. A number that is 36% of its reciprocal
c. A number that is 2.1 greater than its reciprocal

Another aspect to interweaving is retrieval. Retrieval practice is so important for ensuring learners remember what they have studied. This is often done in lessons through starters or low-stakes quizzes, but interweaving can provide a more natural way to incorporate prior learning into lessons. Instead of separating retrieval from the new content being taught, it brings it together, allowing previous topics to be explored in new contexts. For example, learning about averages gives opportunities to recap earlier topics. And vice versa, it is often possible to ask an averages question in the context of a new topic being taught.

Interwoven tasks also require learners to think more deeply about the topics being studied. Questions often allow for multiple different approaches and for finding shortcuts through a deeper mathematical understanding. One of my favourite examples of this is determining which of the fractions $\frac{35}{99}, \frac{35}{100}$ and $\frac{35}{101}$ would round to 0.3 (1 d.p.). Through a combined understanding of fractions and of rounding, converting fractions to decimals can be avoided. Similarly, 'unsimplifying' $3\sqrt{11}$ can help decide if it would round to 10 to one significant figure. This flexibility of understanding is so valuable for problem solving and going beyond simply following procedures.

Questions featuring interweaving are very challenging, often suprisingly so. This level of challenge can cause difficulties, especially if the tasks aren't used well. But it also gives opportunities for stretching learners of all abilities, rather than just accelerating them through the curriculum. Interwoven tasks are also easier to ramp up in difficulty, as there are more moving parts that can be adjusted. For example, equations with numbers in standard form can be made more challenging by adapting the equations or by adapting the numbers or by a combination of both, such as: Solve $23x + 4 \times 10^5 = 5 \times 10^6$.

Interweaving can also help give a purpose to new topics being studied. Learners can see how their new skills can be applied in familiar contexts; e.g. learning about trigonometry can unlock geometry questions that could not previously be solved or give alternative methods for ones that could already be answered. I like to show a Pythagoras question and ask students to answer it as if they had forgotten Pythagoras' Theorem.

Increasingly, GCSE maths exams are including interwoven questions. In fact, the infamous 'Hannah's sweets' question is a prime example. The more students encounter interwoven questions during their journey through school mathematics, the more prepared they will be to tackle such questions when they inevitably arise in examinations.

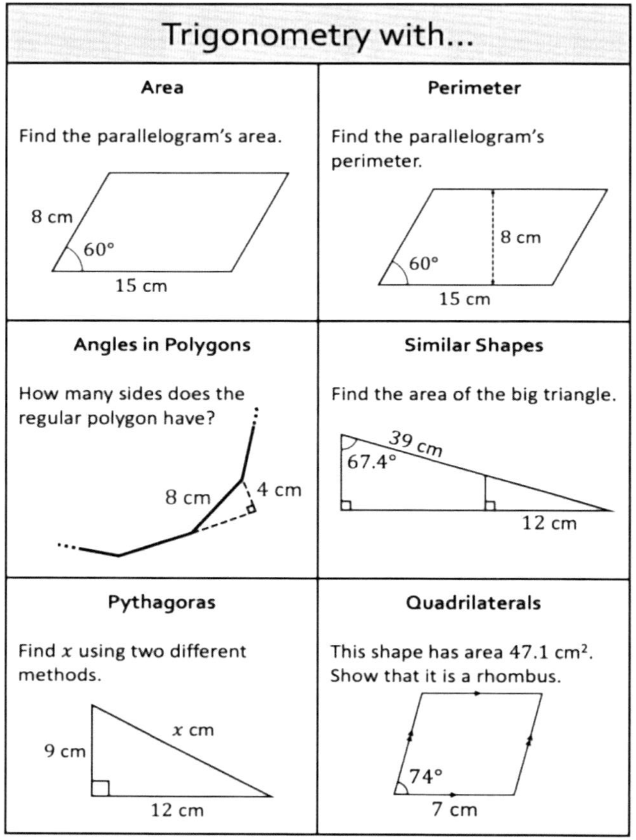

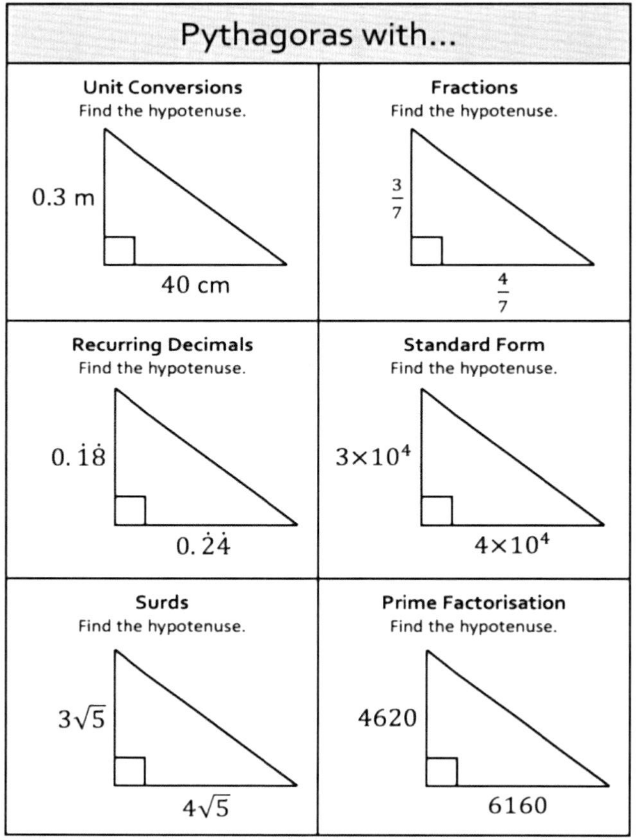

Pythagoras with...

Unit Conversions
Find the hypotenuse.
0.3 m, 40 cm

Fractions
Find the hypotenuse.
$\frac{3}{7}$, $\frac{4}{7}$

Recurring Decimals
Find the hypotenuse.
$0.\dot{1}\dot{8}$, $0.\dot{2}\dot{4}$

Standard Form
Find the hypotenuse.
3×10^4, 4×10^4

Surds
Find the hypotenuse.
$3\sqrt{5}$, $4\sqrt{5}$

Prime Factorisation
Find the hypotenuse.
4620, 6160

Trigonometry with...

Area
Find the parallelogram's area.
8 cm, 60°, 15 cm

Perimeter
Find the parallelogram's perimeter.
8 cm, 60°, 15 cm

Angles in Polygons
How many sides does the regular polygon have?
8 cm, 4 cm

Similar Shapes
Find the area of the big triangle.
39 cm, 67.4°, 12 cm

Pythagoras
Find x using two different methods.
9 cm, x cm, 12 cm

Quadrilaterals
This shape has area 47.1 cm². Show that it is a rhombus.
74°, 7 cm

However, using interwoven tasks in the classroom effectively requires some consideration and some practice. I have found them to be most helpful towards the end of a learning episode, after having had some practice with more standard questions on a new topic. Interwoven tasks are also very helpful for revision, allowing multiple topics to be recapped at once. And while they work well as extension tasks or with high-attaining students, I have found that all my classes have benefitted from getting the time to work together on interwoven problems. Much of the use of these tasks will depend on the class and the task itself. Sometimes it can be best to work through an example first, other times to just give the questions and see what students come up with. Sometimes working through problems one at a time as a class is beneficial, other times students need to work through them at their own pace.

The main challenge when using interwoven tasks is how difficult learners can find them! To overcome this, I find it helpful to have enough questions ready so that learners can get over the novelty of the questions and focus on the maths itself. Having questions that gently ramp in difficulty also helps, starting with the very simplest example possible if necessary. Perhaps recap the topic to be 'interwoven in' separately first, and only interweave the topics together when students are confident with both individually. Where needed, give structures such as backward-faded examples and carefully model how to lay out answers that may be longer than students are used to.

While there are more and more interwoven maths tasks being shared online, it is often helpful to create your own. This way they can be tailored to your scheme of work and the needs of your students. I have also found creating my own interwoven tasks to be a really rewarding experience, helping me think about familiar maths in new ways. The first step to creating an interwoven task is deciding which topics to interweave. I find it helpful to think about three elements to a maths question, each of which can come from different topics: inputs, processes, and contexts.

Changing the form of the numbers in a question is often the easiest way to interweave in other topics. Instead of having the length of a triangle be a whole number, have it be a recurring decimal, or a fraction, or a surd, or in standard form, etc. Changing the inputs like this makes all the calculations in the question less routine. Sometimes it will be best to choose one type of input, for example a set of questions featuring sequences of surds. However, mixing multiple types of input can also be helpful in revealing connections; e.g. in the panel on the previous page, all the triangles are 3-4-5 triangles, some more obviously than others.

Angles in polygons with...

Ratio	Simultaneous Equations
A regular polygon has interior and exterior angles in the ratio $5:1$ How many sides does it have?	A regular polygon's interior angles are 120° bigger than its exterior angles. How many sides does it have?
Percentages	**Averages**
A regular polygon has exterior angles that are 2.5% of the size of the sum of its interior angles. How many sides does it have?	A polygon has one right angle. The mean of its other angles is 150°. How many sides does it have?
Bounds	**Sequences**
A regular polygon has interior angles that round to 150° to 2 significant figures. How many sides could it have?	A polygon has angles that form an arithmetic sequence. Its smallest angle is 135° and its largest angle is 177°. How many sides does it have?

Linear equations from...

Straight Line Graphs	Perimeter
The point $(x, 180)$ lies on the line $y = 2x + 144$. Find the value of x?	Form and solve an equation to find the value of x. $72 \quad \boxed{P = 180} \quad x$
Angles	**Probability**
Form and solve an equation to find the value of x. $144°$ x x	A bag contains x red counters and x blue counters. The rest of the 180 counters are green. The probability of choosing a green counter at random is 80%. Find x.
Functions	**Ratio**
$f(x) = 180 - ax$ When 2 is inputted into the function, the output is 144. Find the value of a.	Anne and Bob share £180 in the ratio $72 : x$ Anne receives £144. Form and solve an equation to find the value of x.

The process of a question is usually its primary topic; it is the maths that must be done to answer the question. For example, the first set of questions on the previous page all require the same process of solving a linear equation, but in different contexts. Other processes could include sharing in a ratio or finding the nth term rule of a sequence. Key processes like these can be incorporated into a vast array of other topics, each time requiring learners to think more deeply about those other topics while getting extra practice of the core process.

The context of a question possibly has the biggest impact of all, as it determines the first impression when seeing the problem. Two mathematically identical questions can seem very different if the context is changed. The context of a question is usually where the necessary factual knowledge is found. For example, the other set of questions on the previous page all require knowing the same key facts about angles in polygons. However, actually solving each of the questions involves a range of different processes from sharing in a ratio to solving quadratic equations.

There is so much room for creativity when shaking up those three aspects to a maths question. I really enjoy coming up with questions that have surprising combinations, such as a sequences question that has inputs in standard form, requires the process of solving an equation, and is in the context of rounding:

> *The first two terms of an arithmetic sequence are 4×10^{-3} and 4.4×10^{-3}.*
>
> *Find the position of the first term in the sequence that rounds to 3 to the nearest whole number.*

Or an averages question that requires dividing surds in a ratio:

> *An irrational number is shared in the ratio $2:3:7$.*
>
> *Given that the mean share is $\sqrt{128}$, what is the size of the smallest share?*

Most of the tasks I've talked about so far involving 'interweaving in' different topics. However, it is also possible to 'interweave out' – finding key ideas that appear naturally across multiple topics and exploring those key ideas in depth. I have found these tasks really helpful for that main aim of making connections, as the connections are already there, and the tasks merely reveal them. One such key idea is proportional relationships, which are prevalent throughout mathematics. A task such as the one on the next page draws attention to that common underlying

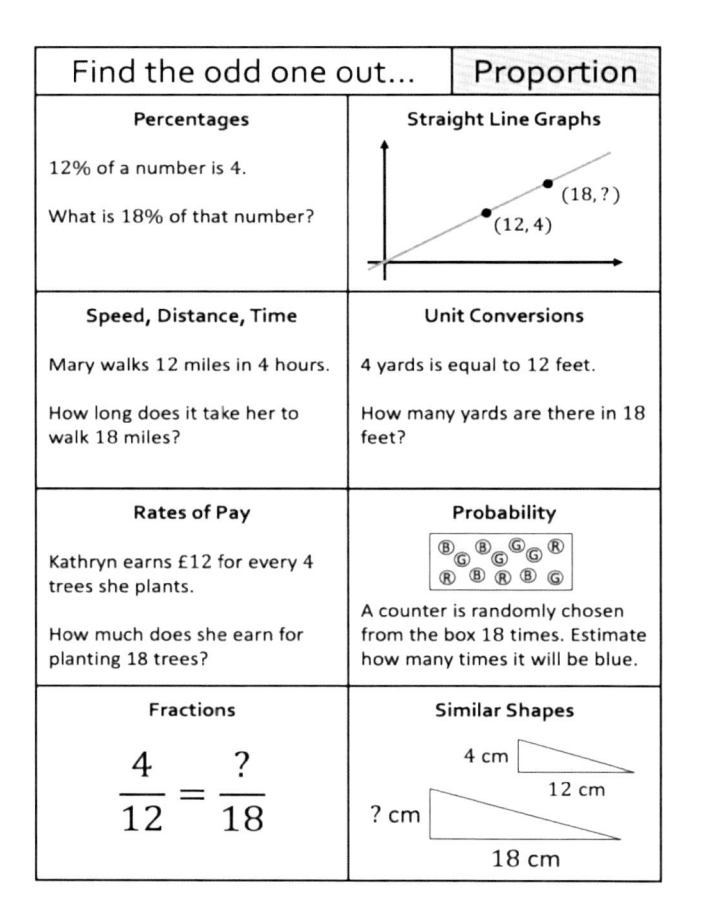

Change each question to make the answer 43

Find the 10th term of the arithmetic sequence starting 7, 10, …	*(straight line graph through (1,7), (2,10), (10,?))*
A taxi fare includes a call-out fee and a price per mile driven. 1 mile → £7 total 2 miles → £10 total 10 miles → ?	1 → → → 7 2 → × → + → 10 10 → → → ?
△ □ = 7 △△ □ = 10 △△△△△△△△△△ □ = ?	*(cups: 7 cm, 10 cm, ? cm)*

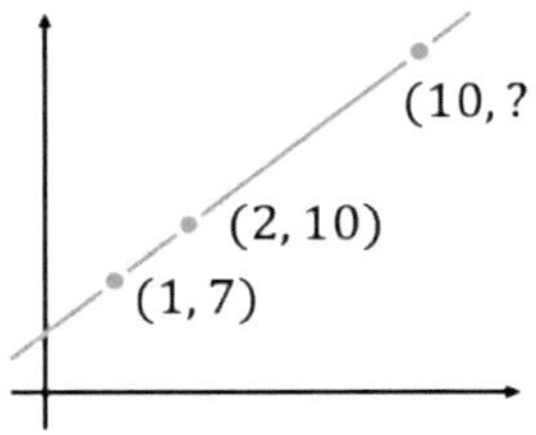

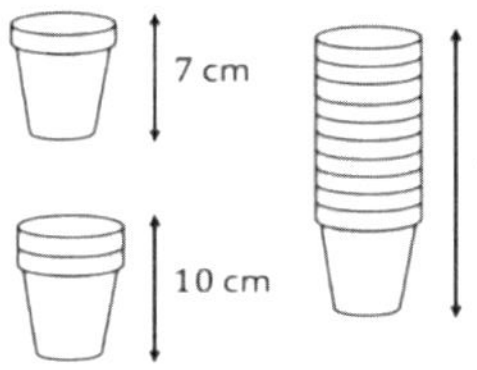

Find the odd one out… | Proportion

Percentages 12% of a number is 4. What is 18% of that number?	**Straight Line Graphs** *(line through (12,4) and (18,?))*
Speed, Distance, Time Mary walks 12 miles in 4 hours. How long does it take her to walk 18 miles?	**Unit Conversions** 4 yards is equal to 12 feet. How many yards are there in 18 feet?
Rates of Pay Kathryn earns £12 for every 4 trees she plants. How much does she earn for planting 18 trees?	**Probability** B B G G R R B R B G A counter is randomly chosen from the box 18 times. Estimate how many times it will be blue.
Fractions $$\frac{4}{12} = \frac{?}{18}$$	**Similar Shapes** *(triangles: 4 cm, 12 cm; ? cm, 18 cm)*

structure, featuring eight contexts for the same mathematical relationship. With tasks like these, working out the answers to the questions is not the purpose. The purpose is to explore the connections between the questions. Not just recognising that the answers are all the same but understanding why they are all the same. It can be helpful to have a prompt that helps learners engage with the questions in this way. One such prompt is to have an 'odd one out' that can be corrected to match with the others.

The first set of questions on the next page all involve linear relationships. Each question includes the number 10 twice. This means close attention must be paid when adapting the questions as to which 10 is being changed. In fact, there are two natural ways to adapt each question, changing the 10 miles (or equivalent) to 13 miles, or changing the £10 (or equivalent) to £11. I like how these questions give meaning to the idea of the 0th term in sequences, which is also the y-intercept for the graph, and the value of the square, etc. And the same for the common difference, which is the gradient of the graph, the price per mile, the height of the rim of the cup, and so on.

This format can be also used to show a range of representations. In this case, there are six different presentations of three equations that must be paired up. This can be followed up by completing blank grids with six similar presentations of a fourth equation. This can also help explore the limitations of each representation by asking which ones would be suitable for showing $2x + 4 = 4x + 10$ or $10 - 2x = 4$.

Sometimes the key idea can appear for less obvious reasons. For example, the other set of questions on the next page all have answers that include reciprocal pairs. With some questions this can be easily understood by thinking of a reciprocal as a multiplicative inverse. Others are less obvious, for example the solutions to $10x^2 - 29x + 10 = 0$, but still not a coincidence. This can lead to a brief investigation of which quadratics have solutions that are reciprocals of one another.

Once the key idea has been thought of, tasks like these are very quick to make and are infinitely adaptable. I have really enjoyed the discussions that have come from using these tasks, with classes often finding things I hadn't noticed. I love how visually diverse the questions can look on the surface, despite them having some underlying hidden connection. And that is what interweaving is all about, revealing the connections that are always lying somewhere beneath the surface.

Pair them up! | Equations

Algebraic

Solve:

$$2x + 4 = 10$$

Worded

I think of a number, add 4, and multiply by 2.

I end up with 10. What did I start with?

Function Machines

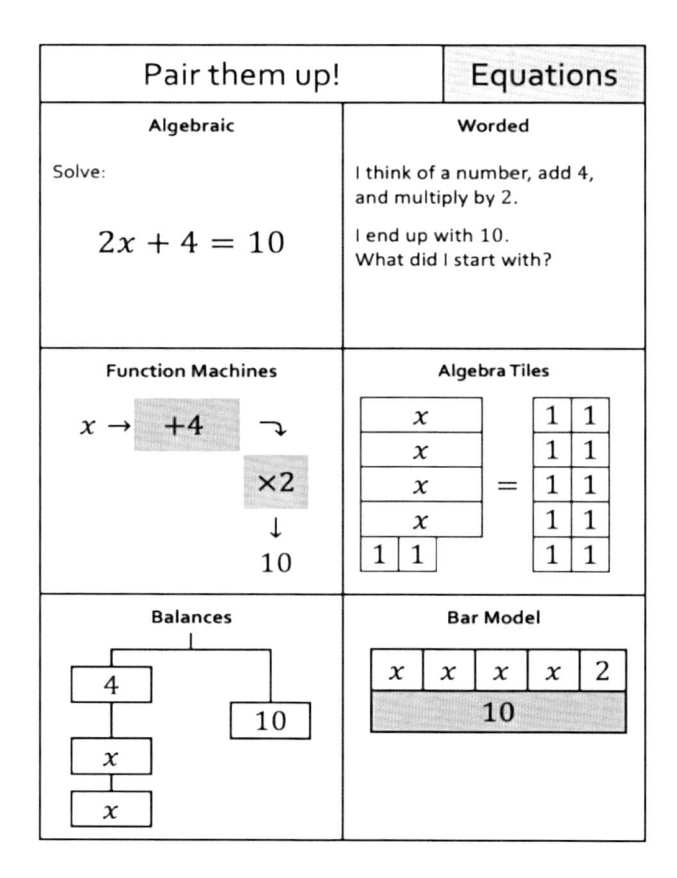

$x \rightarrow \boxed{+4} \rightarrow \boxed{\times 2} \rightarrow 10$

Algebra Tiles

(tiles showing x, x, x, x on one side equals tiles of 1s on the other)

Balances

(balance diagram with 4, x, x on one side and 10 on the other)

Bar Model

x	x	x	x	2
10

What do you notice? | Reciprocals

Ratio

Write 15 : 6 in the form:

a) 1 : n
b) n : 1

Quadratic Equations

Solve:

$$10x^2 - 29x + 10 = 0$$

Speed, Distance, Time

I walk 15 miles in 6 hours.

a) How long does it take me to walk a mile?
b) How many miles do I walk in one hour?

Transformations

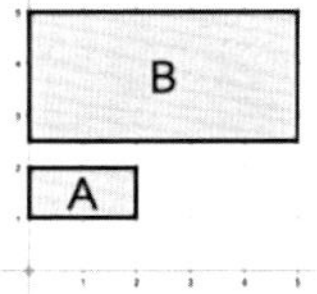

(diagram showing shapes A and B)

a) Transformation from A to B?
b) Transformation from B to A?

Similar Shapes

Cuboids A and B are similar. A has a volume of 8 cm³ and a width of 1 cm.
B has a volume of 125 cm³ and height of 1 cm.
Find the height of A and the width of B.

Trigonometry

A right-angled triangle has shorter sides of length 6 cm and 15 cm, and an acute angle of θ.

What are the possible values of $\tan(\theta)$?

15 Thoughts on A level Integration

Tom Bennison

Since I started teaching, I have been lucky enough to have many conversations with mathematicians and educators, conversations that have really enriched my teaching of A level Mathematics and Further Mathematics (see Note). It is by such fertilisation of ideas through discussion that my teaching has improved with experience. So if I could tell you one, two or three things in what follows, I hope to provide a bit of a "jump start" to those new to teaching integration.

Integration is, on balance, my favourite A level mathematics topic as it is a topic where you can be truly creative; it doesn't feel like you have to just follow rules. It is, however, also a topic that students typically struggle with as experience and mathematical intuition are such an important factor.

Sequencing the teaching of integration

In case you aren't familiar with the integration content of the current mathematics A level, it is reproduced below, taken from the DfE Subject Content guidance, (DfE, 2016). Content for AS level is shown in bold.

Code	Subject Content
H1	**Know and use the Fundamental Theorem of Calculus**
H2	**Integrate x^n (excluding $n \neq 1$), and related sums, differences and constant multiples.** Integrate $e^{kx}, \frac{1}{x}, \sin(kx), \cos(kx)$ and related sums, differences and constant multiples.
H3	**Evaluate definite integrals; use a definite integral to find the area under a curve** and the area between curves.
H4	Understand and use integration as the limit of a sum
H5	Carry out simple cases of integration by substitution and integration by parts; understand these methods as the inverse processes of the chain and product rules

Code	Subject Content
	respectively. (Integration by substitution includes finding a suitable substitution and is limited to cases where one substitution will lead to a function which can be integrated; integration by parts includes more than one application of the method but excludes reduction formulae).
H6	Integrate using partial fractions that are linear in the denominator.
H7	Evaluate the analytical solution of simple first order differential equations with separable variables, including finding particular solutions. (Separation of variables may require factorisation involving a common factor).
H8	Interpret the solution of a differential equation in the context of solving a problem, including identifying limitations of the solution; includes links to kinematics.
I3	Understand and use numerical integration of functions, including the use of the trapezium rule and estimating the approximate area under a curve and limits that it must lie between.

Integration is typically first introduced during Year 12, often after students are familiar with differentiation of $y = x^n$. Integration as the inverse operation to differentiation is often introduced first with its other characterisation as the area under a curve coming a bit later. The teaching of the topic then often proceeds in much the order outlined above – interspersed with other topics – until the topic is completed at some point during Year 13.

I always found it strange that we taught differentiation from first principles but understanding integration as a limit of a sum was not afforded the same prominence. I sequenced integration in the traditional way in my schemes of work, splitting the content between modules as in the previous A level syllabus. With the linear A level we have much more freedom in how we introduce and order topics and this is something we can use to our advantage.

I began to question the status quo as I became more experienced as a teacher, and especially after reading a recent article by Colin Foster and his co-authors (Foster et al, 2021). Through this article I was introduced to Dietiker's notion of the mathematics curriculum as a story with mathematical characters, actions, settings and plot. As a result, I sequenced the integration content differently, as shown in Figure 1, emphasizing it as a method for finding areas first and foremost, before relating integration and differentiation. (If you have a planimeter you could even talk about the history of finding areas of shapes before starting the mathematics.)

In order for students to be able to integrate from first principles, it is necessary to move the work on arithmetic and geometric series before integration, a departure from studying this content in the second year of the course. Following the sequencing diagram shown in the panel below, we shall take a deeper look at integration from first principles.

Approximating the area under a curve	
Finding areas by the trapezium rule	
Integration as the limit of a sum	Riemann integration
Finding the integral of x^n (excluding $n \neq 1$) by first principles	
Know and use the fundamental theorem of calculus	
Find the area under a curve analytically	Find the area between curves analytically
Integration of standard functions (e^{kx}, $\frac{1}{x}$, $\sin(kx)$ and $\cos(kx)$) including the use of first principles	
Methods of integration: Reverse chain rule Integration by parts Integration by substitution Integration using partial fractions	
Area under curves defined parametrically	
Forming and solving simple differential equations	
Separation of variables	

Integration from first principles

We consider the problem of finding the area between $f(x)$, the x-axis and the two limits $x = a$ and $x = b$.

We begin by dividing the interval $[a, b]$ into n subintervals by defining the distinct points
$$a = x_0, x_1, \ldots, x_{n-1}, x_n = b$$

Let $h_i = x_i - x_{i-1}$, $i = 1, \ldots, n$ and define the midpoint of each subinterval as
$$\overline{x_i} = \frac{x_{i-1} + x_i}{2}, \quad \text{for } i = 1, \ldots, n$$

We can then construct rectangles of width h_i and height $f(\overline{x_i})$.

The area of each rectangle is then $hf(\overline{x_i})$ and the total area under the curve can be approximated by summing the area of all the rectangles, that is $A \approx \sum_{i=1}^{n} h_i f(\overline{x_i})$.

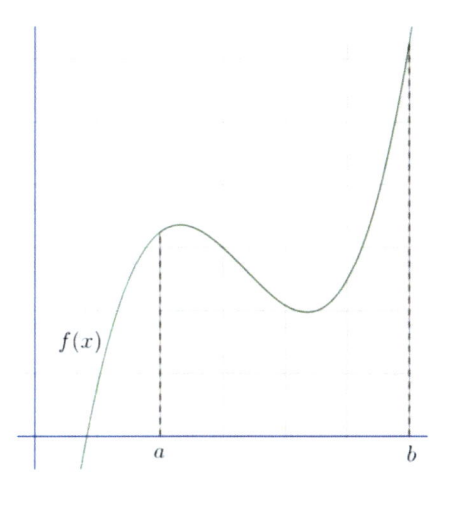

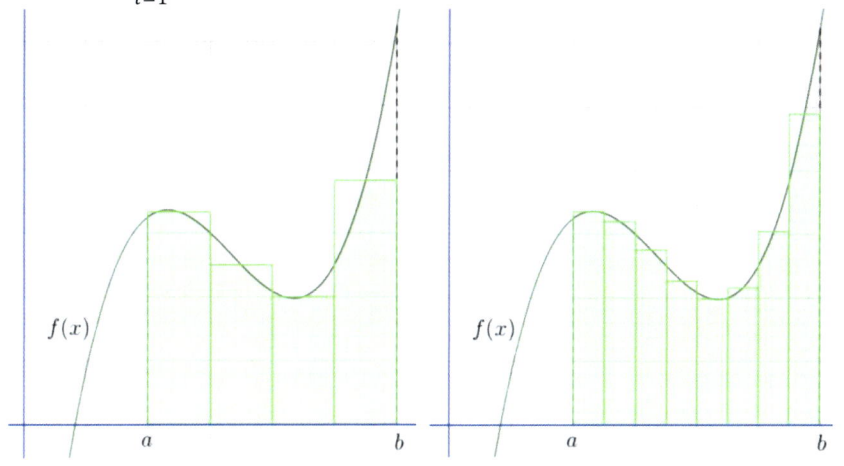

For simplicity let us assume that $h_i = h$ for all i, then, as we increase the number of intervals n (and so reduce h), we can expect the value of A to converge to the area under the curve.

We are now in a position to define the Riemann integral of $f(x)$ by the definition

$$\int_a^b f(x)dx = \lim_{n\to\infty} \sum_{i=1}^n h_i f(\overline{x_i})$$
$$= \lim_{n\to\infty} \sum_{i=1}^n h f(\overline{x_i})$$
$$= \frac{b-a}{n} \sum_{i=1}^n f(\overline{x_i}).$$

This formula can then be used to find the areas under a range of standard functions. It is sensible to begin with a function that has an area they can already find.

Example: $f(x) = x$

Suppose we wish to find the area under the straight line $f(x) = x$ with the limits $x = a$ and $x = b$ then we know that we can find the area by a standard application of the formula for the area of a trapezium. Hence, we already know that the area we seek has value,

$$A = \frac{a+b}{2} \times (b-a) = \frac{b^2 - a^2}{2}$$

Let us pick n subintervals, all of the same width h, where $h = \frac{b-a}{n}$ and define $\overline{x_i} = a - \frac{h}{2} + ih$, for $i = 1, \ldots, n$. Then, working with the definition we obtain

$$A = \lim_{n\to\infty} \sum_{i=1}^n h f(\overline{x_i})$$
$$= \lim_{n\to\infty} \sum_{i=1}^n \frac{b-a}{n}\left(a - \frac{h}{2} + ih\right)$$
$$= \lim_{n\to\infty} \frac{b-a}{n} \sum_{i=1}^n \left(a - \frac{h}{2} + ih\right).$$

Now, considering the terms in the summation we notice that they form an arithmetic progression, with initial term $a + \frac{h}{2}$ and common difference h. The final term is then $b - \frac{h}{2}$.

Using the formula for the sum of an arithmetic progression we have,

$$\sum_{i=1}^n \left(a - \frac{h}{2} + ih\right) = \frac{n}{2}\left[\left(a + \frac{h}{2}\right) + \left(b - \frac{h}{2}\right)\right] = \frac{n}{2}(a+b).$$

Hence,
$$A = \lim_{n\to\infty} \frac{b-a}{n} \sum_{i=1}^{n} \left(a - \frac{h}{2} + ih\right)$$
$$= \lim_{n\to\infty} \left(\frac{b-a}{2}\right)(a+b)$$
$$= \left(\frac{b-a}{2}\right)(a+b)$$
$$= \frac{b^2 - a^2}{2}, \quad \text{as we expected.}$$

Further examples are, perhaps, more involved but strong students will enjoy the challenge of computing from first principles the area under $f(x) = x^2$, $f(x) = \frac{1}{x}$ and $f(x) = \sin x$.

Following this we can relate integration and differentiation by considering the following. Let $A(x)$ be the area under the curve $f(x)$ between $x = 0$ and $x = x$. We now consider the area under the graph between x and $x + h$, where h is some small increment, as shown in the diagram below.

Let $f_{\min}$ be the smallest value that $f(x)$ takes between x and $x + h$ and $f_{\max}$ be the largest value that $f(x)$ takes between x and $x + h$.

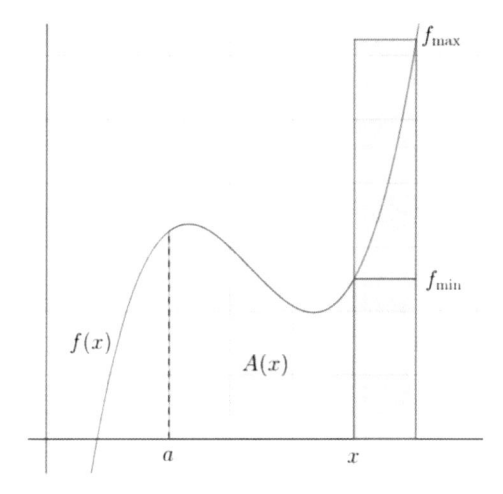

From the diagram it is evident that the following holds,
$$A(x) + f_{\min}h \leq A(x+h) \leq A(x) + f_{\max}h.$$
With a little rearranging, we obtain, $f_{\min} \leq \frac{A(x+h) - A(x)}{h} \leq f_{\max}$.

As $h \to 0$ then both $f_{\min} \to f(x)$ and $f_{\max} \to f(x)$. So we have,
$$f(x) \le \lim_{h \to \infty} \frac{A(x+h)-A(x)}{h} \le f(x).$$

Hence, we can conclude that $\frac{dA}{dx}(x) = f(x)$, demonstrating that differentiation is the reverse of integration.

A fun integral

This is an interesting integral to explore with further mathematicians, though it could also be done with students who don't do further mathematics if you provide them with the formula book.

Consider the integral
$$I = \int_0^1 \frac{x^4(1-x)^4}{1+x^2}.$$

With some questioning we can guide our students to determine if this integral is positive or negative and have some appreciation of its possible size.

We can now ask them to evaluate the integral; they will need to expand some brackets (good practice of the binomial) and then perform polynomial division.

$$\frac{x^4(1-x)^4}{1+x^2} = \frac{x^4(x^4 - 4x^3 + 6x^2 - 4x + 1)}{1+x^2}$$
$$= x^6 - 4x^5 + 5x^4 - 4x^2 + 4 - \frac{4}{1+x^2}.$$

Hence,
$$I = \int_0^1 \left(x^6 - 4x^5 + 5x^4 - 4x^2 + 4 - \frac{4}{1+x^2}\right) dx$$
$$= \left[\frac{x^7}{2} - \frac{2}{3}x^6 + x^5 - \frac{4}{3}x^2 + 4x - 4\tan^{-1} x\right]_0^1$$
$$= \frac{22}{7} - \pi$$

which is a very small positive number. This demonstrates that $\frac{22}{7}$ is a fairly good approximation to π, but it is an overestimate.

Reverse Chain Rule versus Integration by Substitution

From experience I know that there are some strong views on this topic! Over time, I have become more convinced of the importance of teaching integration by inspection (i.e. using the chain rule in reverse) before teaching the formal method of integration by substitution.

There is of course the saving in time if students can recognise integrals such as $\int 3x(2x^2 + 5)^7 dx$ as opposed to going through the integration by substitution procedure with $u = 2x^2 + 5$. With practice though students can become remarkably quick performing such a substitution.

Of greater importance is the impact on the perception and psychology of the students. Encouraging students to try to spot the answer and adjust before resorting to an established method encourages creativity in the students, emphasising that integration is more than a sequence of methods to follow. It requires practice, intuition and dexterity with a range of approaches; all of this makes integration deeply satisfying.

The importance of multiple methods

Integrals can often be evaluated in multiple ways. Back in 2019 I was inspired to make a resource for one particular integral having got involved in a conversation on Twitter with Mike Lawler (@mikeandallie) and some others.

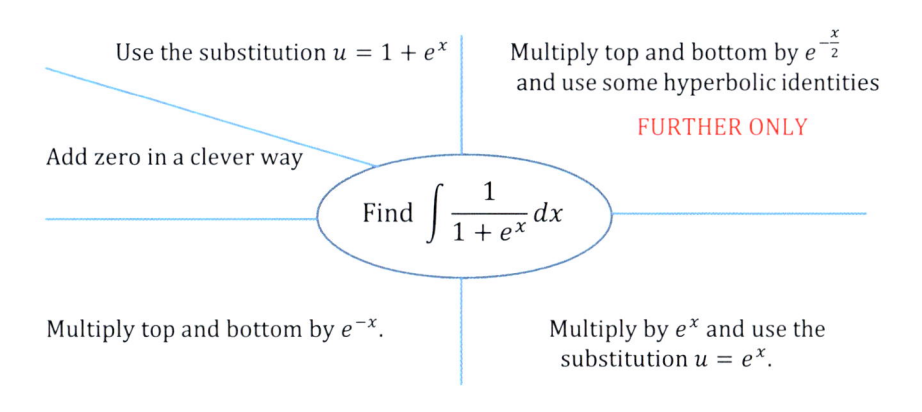

Find $\int \dfrac{1}{1 + e^x} dx$

- Use the substitution $u = 1 + e^x$
- Multiply top and bottom by $e^{-\frac{x}{2}}$ and use some hyperbolic identities. **FURTHER ONLY**
- Add zero in a clever way
- Multiply top and bottom by e^{-x}.
- Multiply by e^x and use the substitution $u = e^x$.

References

Department for Education, 2016. *Mathematics AS and A Level Content*, https://assets.publishing.service.gov.uk/government/uploads/system/uploads/attachment_data/file/516949/GCE_AS_and_A_level_subject_content_for_mathematics_with_appendices.pdf

Foster, C., Francome, T., Hewitt, D. & Shore, C. 2021. 'Principles for the design of a fully-resourced, coherent, research-informed school mathematics curriculum', *Journal of Curriculum Studies* 53(5), 621-641.

Note

I'd like to thank Edward Hall in particular as some of the ideas discussed here were originally part of a Mathsconf workshop we delivered together. I'd also like to mention Susan Whitehouse, Luciano Rila, David Bedford, Mike Lawler, Jim Hardy and Colin Foster with whom I've had memorable conversations on the topic of calculus over the years.

16   *When am I Ever Going to Use This?*

Dave Gale

Is maths useful?

I'm going to start with a potentially controversial view. While I certainly think maths is useful, I'm not convinced it is as essential to everyday life as people make out. Don't get me wrong, I know that there's absolutely loads of maths behind many of the things that we use on a day-to-day basis and things like cars and mobile phones simply wouldn't work without the maths behind the scenes. However the fact that many people leave school with a relatively low GCSE grade and still go on to lead functional lives as adults suggests to me that not all of the things taught in maths at school can possibly be *that* essential. At the risk of sounding like a stereotypical year 9 student, realistically when is a typical person ever going to use quadratics in their day-to-day life? All of the essential maths that goes into making my kettle turn off at the right time is entirely hidden from the user and I don't need to understand or even be aware of that maths to enjoy my coffee. The good news is that I don't believe the main reasons for studying maths are down to its utility. I think we teach people maths because of the way it helps your brain think, for the satisfaction that you can get from solving problems and the beauty that can be found within its many wonderful facets. Now, it's possible that you disagree with this view and that's fine, of course. I would ask you to consider this though, if you like/love maths is it because you adore how wonderfully functional it is or is it for some other reason?

But sir, when am I going to use this?

If you've been teaching maths, you've almost certainly been asked this. It's not a unique question for maths teachers as languages teachers and English teachers get this too while it's a rare question for art teachers. Students are categorically told that they *need* maths to progress to whatever it is they're doing next or for 'later life' so it's understandable that they might want to know where they could find themselves needing surds or bisecting an angle with a pair of compasses. I've answered this question in various ways over my teaching career and here are some of them. You can decide for yourself if you'd like to make use of any of them:

- Well, topic X is used in many areas of engineering. (I never actually knew which areas though.)
- This topic comes up in the thing we're doing next. You'll need this to be able to do that.
- I'm not convinced you really want to know the answer to that but, if you do, come back at break and we can talk about it then. (They almost never do.)
- I don't know but it's on the test, so you'd better learn it.
- You're probably not. It's fun in its own right though and you'll feel a sense of smug, self-satisfaction when you can do it.

Context, the abstract and pseudocontext

At some point in your career, someone will tell you that students engage with work better if the work is relevant to them. I think that can be incredibly difficult for maths teachers and, frankly, a bit misleading. Some maths is useful to some people. Some maths is purely abstract and therein lies its beauty. If the work you're doing with students is going to be better when placed into a context, then I would certainly recommend that. If the work is more abstract in nature, then I recommend keeping it that way. There are some topics where I have tried very hard but have just not found ways to make them relevant to students and even for the ones that I have, it doesn't always mean that they are inherently more interested. If you know a meaningful context for surds that is relevant to teenagers, then please do let me know!

Contextual questions

Maths questions in a meaningful, realistic context are some of the best things to use in a class. The trouble is, they're not that easy to find or create successfully. The favourite lessons I've taught have been using ideas from Level 3 Core Maths in GCSE classrooms. The critical analysis part of the course is perfectly applicable for younger students and asking them to fact check items in the news or on social media is often very engaging. Encouraging students to question numbers that they see in headlines is an excellent use for maths while also helping to develop their social awareness.

Abstract maths

It's not unusual for teachers of maths to lean towards enjoying the algebraic, pure aspects of the subject. I think that a significant part of thinking that you are 'good at maths' comes from being able to manipulate some algebraic items and get to a 'nice' result in some way.

There's an undeniable sense of satisfaction that comes from being able to do something that is perceived as difficult and to some extent, for other people to know that you can do it. There are parallels to being able to juggle or being able to knit well. The chances are that if someone is good at those things they get great satisfaction from being able to accomplish a new technique and, quite possibly, they'll want to share that with someone. I think that this feeling is similar to the one a student gets from completing a puzzle or challenge in maths.

It's not easy to get students to the point where they see maths as being worth doing in its own right so I sometimes talk to them about jigsaw puzzles. Most students have done a jigsaw puzzle at some point and probably got a sense of satisfaction at completing it. But, why were they doing it? It wasn't really to get the 'final answer' because the picture was already on the box. It wasn't to get a beautiful thing to look at and keep as, in most cases, people dismantle the puzzle and put it away pretty soon after doing it. It must have been for the satisfaction of completing a challenge and for the mini-challenges along the way. I argue that that is the same mindset I'd like them to have when approaching maths questions. If students have ever played a computer game, done a sudoku or spent time perfecting a skateboarding move, they have the right mindset already and they can apply the, "Hey, I wonder if I can do this challenge?" to maths too.

Once you've got that playful, "Let's see if we can do this," approach, puzzles don't need to be in a context. An excellent example is Catriona Agg's shape puzzles from Twitter. Puzzles like this are intriguing in their own right and once they're in the right mindset, students enjoy working on them.

What's the area of the circle?

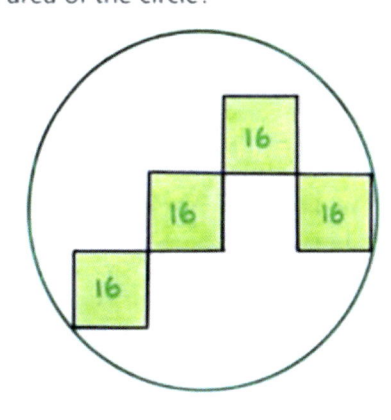

I'm hoping nobody wants to take this problem and turn it into a worded one that starts, "A farmer has four square fields, each with an area of 16." This question doesn't need a context and trying to make one leads us into…

Pseudocontext

Pseudocontext is the reason maths ends up as internet memes. Only in maths class would someone buy 59 watermelons and have nobody question it. As an exercise, consider these two questions regarding forming and solving equations:

1. A plumber charges a £50 call out fee and £40 per hour.
   (a) Write an expression for the total cost, C, for a job that takes h hours.
   One job carried out by this plumber was charged a total of £170.
   (b) Set up and solve an equation to find out how long this job took.

2. A normal train carriage holds d people. Two new types of carriage are being developed. The mini-carriage holds 15 people fewer than the normal carriage, and the maxi-carriage holds 50 people more.
   (a) Write down an expression for the number of people a
       (i) mini-carriage holds, (ii) maxi-carriage holds.
   The maxi carriage holds 120 people. Use this information to form an equation.
   (b) Solve this equation to find out how much a
       (i) normal carriage holds, (ii) maxi-carriage holds.

The first question is mostly reasonable as this is the sort of calculation that the plumber would need to perform to work out how much to charge. I would accept that they may not be thinking of it as a formula but I also think it's plausible that they might want to use a spreadsheet to help them track their jobs. In this case, being able to think of the costs in the format of a formula would help with being able to convert it to a calculation in a spreadsheet. Part b is a bit more of a stretch but just about in the realms of what the customer might want to do. "I know I got charged £170 so how long is the plumber claiming that tap took them to fix?" It's not ideal and does ignore the cost of parts but it is just about within the 'plausibly could occur' realm.

The second question is bizarre. The only way this scenario occurs is if the person writing the question already knows all of the answers and is just choosing to pose it like this to see if you can work it out. It's a puzzle and may as well be set up as such. "I have three cards with numbers on them. The lowest one is 15 less than the middle one…" I'm not saying that this would make it a more engaging question but at least it removes the pretence that this would ever happen. I have no doubt that the train context fails to engage students who aren't interested in trains. Worse still, I also suspect it annoys those students that actually are interested in trains because they can probably tell you the standard sizes of carriages anyway.

I do have one subset of pseudocontext questions that I actually do like to use. Sometimes it is fun to be silly and ask some tongue-in-cheek questions. An example that amused me is from a text book and says, "My favourite stick is 14 cm long. Will it fit on my favourite rectangular tray which measures 7cm by 12cm?" You probably have to judge whether this kind of question would go down well with the class you're teaching but I think it's overtly ridiculous nature gives it a pass for its unrealistic context.

So, why are we learning this?

Maths has its uses. Maths can be engaging when set in contextual situations. Maths can be interesting when taught in abstract ways. So, what do I say to students when they ask me this question now?

"You're learning maths so that you can use it in the times when it's useful. You're also learning it because this is the only subject that will get your brain to think in certain, abstract ways. On top of that, maths has many beautiful aspects to it that you can enjoy for its own sake."

17   *I Wouldn't Tell You Anything*

Charlotte Hawthorne

'I wouldn't tell you anything' would be a short chapter indeed. But when I thought about what I wish I'd known, it occurred to me that many of the things I've learnt about maths teaching in the last ten years wouldn't have been as impactful without the experiences I'd had leading up to them. However, I will pass on a couple of carefully chosen nuggets of wisdom if you can call them that.

My first one would be to have an awareness that you may change your mind. There are many things I've changed my mind about at least once and may even change back again one day. As is the case with many teachers, early in my career I was heavily influenced by what I perceived to be the beliefs and practices of my university tutors and school mentors. I remember being convinced that students should discover as much maths as possible by themselves. If the objective was for students to learn about angles formed by transversals crossing parallel lines, I would definitely have had students measuring and trying to discover alternate and corresponding angles for themselves. Of course, the sorts of things they found out were "Miss, I've found that some angles are one degree more than some other angles", or worse "these four don't add up to 360 degrees even though they're around a point". In other words, we found out about human error when measuring angles. I used to do the same with circle theorems too. "Right everyone, draw some circles…". What would I do now? I'd use a dynamic geometry package like *Geogebra* to demonstrate, or ideally let the students explore if they have access to the technology.

Another classic discovery-style lesson is the one where we would 'discover' π. I used to have lots of circular objects, string, scissors, measuring equipment, maybe a big table drawn on the board so that we could divide the circumference by the diameter, and all realise that some students were simply bad at measuring! So I'd do almost the exact same lesson, with a small twist. I'd tell them about π first, I'd talk about the special ratio between a circle's diameter and its circumference. Then the measuring would be confirmation and would serve the purpose of letting students realise that it is difficult to measure the circumference and that it might be useful to have a quicker way to work it out. Students could take a circumference measured with string and see how it comes out just that bit more than three times the diameter, adding a bit more credibility to the irrational number, π. It might have felt like a waste of learning time but was actually an important step in feeling the maths, in having something real and tangible to refer back to, another way to make sense of the new learning. That's not to say it's the only way to 'make sense' of what π represents and why it just has to be a little more than three. I love the images below:

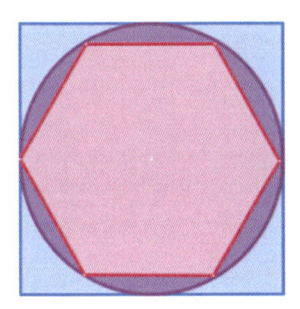

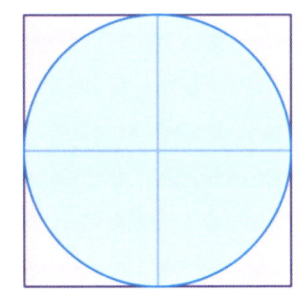

The one on the left is a regular hexagon inscribed in a circle which in turn is inscribed in a square. If the square has unit side length its perimeter is four and by splitting the regular hexagon into equilateral triangles by joining opposite vertices it can be easily shown that the hexagon will have a total perimeter of $6 \times \frac{1}{2} = 3$. The perimeter of the circle must be between three and four. The diagram on the right gives similar insight - we can see that the area of the square is greater than the area of the circle it contains, and the area of the square is four lots of the radius squared. It seems sensible that three of those squares might not account for the whole area of the circle and again it seems that π lots would be reasonable.

I still have debates (often with myself) about what I should lead students to discover or work out and what I should just tell them. I think there's room for both and having a range of approaches at your fingertips instead of just one isn't a bad thing.

My second piece of advice comes in two parts and is something I really do wish I'd known earlier in my career.

1. Zero pairs are amazing; I don't know how I ever taught without mentioning them.
2. Teaching subtraction as addition of additive inverses might sound complicated and take longer to teach but my goodness is it worth it!

Zero pairs are not new. I have a Heylings textbook, published around 1984 on 'Negative numbers and graphs' where positives and negatives are represented as yin and yang parts of a circle which together make zero. Today many teachers, myself included, use two-colour counters where yellow side up represents positive 1 and red side up represents negative 1,

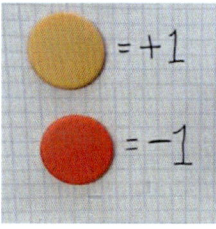

and so two counters showing opposite sides are called a zero pair.

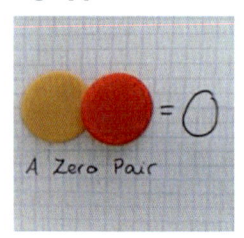

What we are really talking about here are additive inverses. The Additive Inverse Axiom states that the sum of a number and the Additive Inverse of that number is zero:

$$x + (-x) = 0.$$

Using this model any subtraction can be written and modelled as addition of additive inverses. For example,

$3 + (-5) - 7 - (-9)$

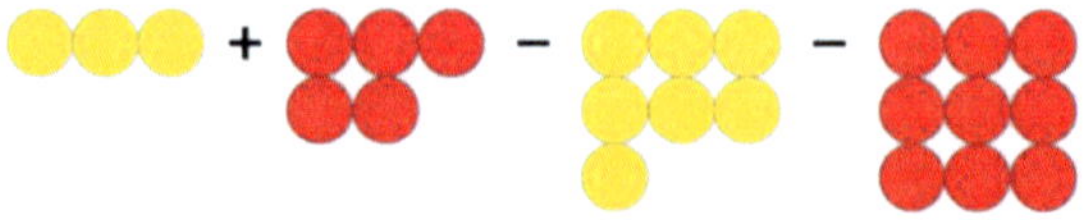

$3 + (-5) + (-7) + 9$

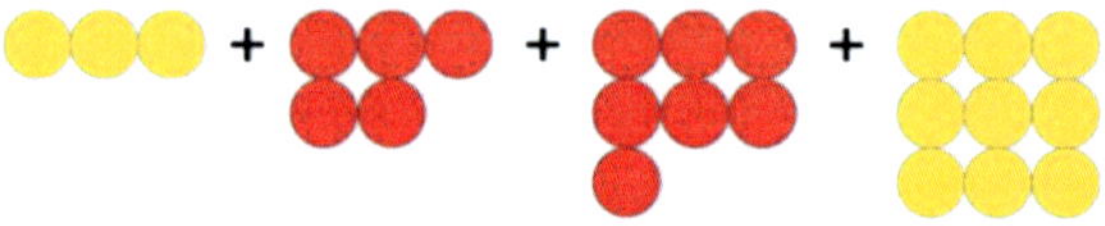

But why bother? Most demonstrations of subtraction of negatives using two-colour counters talk about adding in zero pairs, so why not just do that? Well, it's not only that students find the addition of any number of zero pairs confusing but just how useful I've found addition of additive inverses for other topics.

Here are a couple of my favourites:

Subtraction of vectors

$$\begin{pmatrix} 3 \\ -5 \end{pmatrix} - \begin{pmatrix} 7 \\ -9 \end{pmatrix}$$

Subtraction is addition of the additive inverse (the negative of) the second vector:

$$\begin{pmatrix} 3 \\ -5 \end{pmatrix} + \begin{pmatrix} -7 \\ +9 \end{pmatrix}$$

Simultaneous equations

$$\begin{aligned} 3x - 7y &= 19 \\ -5x - 7y &= -13 \end{aligned}$$

To eliminate y subtract the equations, or add the additive inverse

$$\begin{aligned} 3x - 7y &= 19 \\ +\ 5x + 7y &= 13 \end{aligned}$$

Linear equations with the unknown on both sides

$$3x - 7 = 9 - 7x$$

Although here I wouldn't re-write each subtraction as addition of additive inverses, I would ask what could be added to eliminate the unknown from the right-hand side of the equation, or 'make the zero-pair'.

$$\begin{aligned} 3x - 7 &= 9 - 7x \\ +7x &\qquad\quad +7x \end{aligned}$$

It's surprising how often arithmetic with negatives comes up and having a solid foundation to refer back to is invaluable instead of some misremembered/misunderstood uttering about 'two minuses making a plus' which causes so many more issues than it was ever intended to solve.

Finally, don't forget about the students. Context matters. And it may sound like a cliché but every class really is different. What worked excellently with one class may be far from the best option for another class. Change the way you do things to meet their needs rather than insisting that a particular approach works perfectly, or it's the way you've always done it. Try something new, even if it's not your preferred style, you might just surprise yourself.

So, it was a little more than nothing, but make sure you question everything (even what I've said), read a lot of books and listen to podcasts by people you think will challenge your current thinking. Enjoy the journey!

18 Expect Maths Teachers to Agree on 'Good Learning' but not 'Good Teaching'

Jen Shearman

Introduction

Consider the problem below. It is called 'Make 37' and both the problem and the solution are available online (NRICH, 1997). Enjoy wallowing in some mathematics: take as much time and space as you need to find a solution and convince yourself that you are correct.

> Four bags contain a large number of 1s, 3s, 5s and 7s.
>
> Pick any ten numbers from the bags above so that their total is 37.

Now look back and think about your approach to the problem. Did you:

1. Get out a pencil and paper and try different combinations of the numbers as a 'way in' to solving the problem?

2. Consider carefully the numbers in each bag, sit back in your chair, fold your arms, and smile knowingly?

3. Do something else entirely?

Regardless of your approach you probably found the solution reasonably quickly. I suspect that you probably did not solve the problem by sheer 'brute force' (going through all possible combinations of ten numbers before coming to an answer). At some point your mathematical reasoning about the underlying structure of, and relationships in, the problem will have led you to the solution.

As a mathematician and as a teacher you will encounter broad agreement with others about what a mathematical 'right answer' is. You are, or will become, very adept in spotting which students have a good knowledge, grasp or understanding of a particular mathematical topic.

If I could tell you one thing it is that teachers will disagree strongly about how best to instil this knowledge in their students. It is therefore important for you to have a good understanding of your own beliefs about teaching and learning, and that of the school you work in.

One definition for good learning

In 2019, I asked 45 specialist teachers of mathematics to define *mastery* in relation to teaching and learning in mathematics. These teachers were at different stages of their career, with a variety of training and professional development experiences.

The research found that teachers agreed on what mastering mathematics *means*. My interpretation of their responses defined mastery as 'shared understanding through individual representation'. The teachers felt that their students achieved mastery by acquiring topic-specific knowledge, with problem-solving developing as a result. Students used their own mathematical representations to support their understanding as it developed. One research participant summed this up as:

> I understand teaching for mastery to mean teaching for deep and sustainable understanding; knowing why as well as knowing what and how.

(At least) four viewpoints for good teaching

Whilst teachers agreed on what mastery was and how they would recognise it, what they fundamentally disagreed on was *how* to teach students to acquire mastery. Four different opinions emerged (see Table 1), and these were classified according to teachers' beliefs about their students' capability to master mathematics and how much direction the teacher should give their students.

		Belief about students' capability	
		All can master	Only some will master
Belief about teachers' role	Students to choose representation	Viewpoint 1: 'Travel far, travel together'.	Viewpoint 2: 'Know your limits, follow the teacher.'
	Teacher to choose representation	Viewpoint 3: 'Create a curriculum for interconnected understanding'.	Viewpoint 4: 'Variety in teaching, learning and achievement'.

Table 1: Four viewpoints of good mathematics teaching

Four ways to 'Make 37'

These four distinct viewpoints teachers held about the 'best' way to teach mean big variations in classroom practice. In my time as a classroom teacher, teacher educator and professional development lead, I have been privileged to observe lessons taught by teachers holding all four viewpoints (and more). To illustrate this, I used the findings from my research to create vignettes describing how each type of teacher might approach the 'Make 37' task in the classroom.

Viewpoint 1: 'Travel far, travel together'

Teaching: Teachers aligning with Viewpoint 1 believe all students capable of tackling the problem, and place students in mixed prior-attainment groups. After posing the problem, they give out a variety of resources to help the students, perhaps multilink cubes. The groups are left for a time to contemplate the problem and how to represent it using the resources. Students are encouraged to try out ideas on paper, using the resources and talking to each other. They use their professional judgement when deciding when and how to intervene. They ask students to describe their progress and help students notice important aspects of their activity, such as what the numbers 1, 3, 5 and 7 have in common and what happens when '10 lots of the numbers' are added together.

Learning: The teacher values all mathematically relevant student insights and is not preoccupied with how many of the students solve the problem so long as they all engage in appropriate mathematical thinking which furthers their knowledge.

Viewpoint 2: 'Know your limits, follow the teacher'

Teaching: Teachers aligning with Viewpoint 2 make a judgement about which students can solve the problem and place the students in groups of similar attainment. Lower-attaining students are given something simpler to do. They recognise that fluency in times tables will make this problem easier to solve, so start the lesson with some whole-class times table practice. They display the problem for the class without any additional resources. The teacher asks the class, "What is the highest possible total?" and "What is the lowest possible total?" The teacher then tells the students to draw a table in their books (with columns labelled numbers used and total) and asks students to try different combinations of numbers, suggesting they work systematically and individually. After a length of time the teacher asks how many students have solved the problem. The teacher writes on the board, 'all pairs of odd numbers sum to an even number' and proves it using algebra. Students copy the statement and the proof into their books.

Learning: Students who solve the problem correctly are praised. The students who understand the proof have mastered the mathematics in the lesson.

Viewpoint 3: 'Create a curriculum for interconnected understanding'

Teaching: Teachers aligning with Viewpoint 3 believe that by the end of the lesson all students will be able to solve the problem if they plan the lesson properly, and place the students in mixed-attainment groups. The focus of the lesson is understanding the nature of odd numbers and their multiples. The teacher starts by discussing the nature of '3'. The class recall the 3 times table and the teacher draws attention to the odd and even multiples of 3 using different visual images on the board and multilink cubes. The teacher puts the images of the bags on the board (but not the question) and asks the students to sum any ten numbers from the bags. The teacher strategically selects some of the students to explain their calculation, then poses the 'make 37' problem and asks students to work on the problem individually.

Learning: Most of the class demonstrate that the problem is impossible with different levels of sophistication. Students who understand that pairs of odd numbers will always be an even number have mastered the lesson content.

Viewpoint 4: 'Variety in teaching, learning and achievement'

Teaching: Teachers aligning with Viewpoint 4 believe that all students should be able to attempt the problem if they want to, but not all will be able to solve it in the limited time available. The students are placed in ability groups. They set 'Make 37' as a task that can be attempted by students after they have completed some textbook questions. They introduce the task briefly then devote their time to individual students with whichever activities they are working on. Some groups of students solve the problem and discuss their reasoning. The teacher notices that about six students are stuck on the problem and calls them over to a table to discuss the problem together. No students have found an answer and one student asks if the problem is impossible. The teacher directs the question back to the group of students, who notice that all the totals they have worked out are even. One of the students suggests that all possible totals must be even, and between them, the teacher and the student group work out the reason for this.

Learning: The groups of students have mastered something about the structure of odd numbers through discussing the problem.

What this should tell you

Being a professional is about remaining academically critical. Maintain a balance of scepticism and open-mindedness about how to teach mathematics. You will have had strong emotional reactions to all four approaches above: how can you make use of alternative viewpoints to make you a better teacher?

Expect, and embrace, disagreement

All teachers' beliefs about effective teaching are deep-seated and rooted in a complex web of previous experiences and educational ideologies. And so they should be: we are professionals and have an obligation to make the best decisions we can about how to educate our students. But the different evidence bases we use to inform our decision-making lead to vastly different choices in the classroom and this has implications for us as individuals, the institutions we work in and the profession we are proud to be a part of.

Allow your beliefs to develop and change

You probably agree with your colleagues about what makes a capable and competent mathematician. However, the research above found that there is more that divides mathematics teachers than unites them: the way in which you prefer to teach mathematics will depend on your beliefs about student capability and the role of the teacher in leading students' learning and this may differ from your peers, colleagues and senior leadership.

These beliefs can, and should be, shaped and honed through critical reflection and engagement with all forms of professional development. This includes your own experiences of doing mathematics, your teacher training, collaboration with colleagues, formal development programmes and engagement with both seminal and contemporary educational research. You have a professional responsibility, as do all teachers, to review and interpret the best available evidence about curriculum and lesson planning, classroom practice, school and educational policies and use of resources. You should embrace disagreement and complexity in research.

Understand the beliefs of others before enacting change

If you are in a position of responsibility for mathematics in your school, the idea that 'good teaching' does not share a common philosophy or vision and can mean very different things for different teachers is important to consider. Leaders of mathematics in schools need to understand the individual opinions of colleagues and consider whether and how to change the beliefs and approaches of teachers where they run counter to the department's vision and policy. School mathematics departments and teams include qualified teachers with a variety of mathematical expertise and teaching experience, and pre-service teachers undergoing training. Some department members are likely to be opposed to any specific teaching approach.

In conclusion, the one thing to realise is that there will never be agreement on one way of teaching and this makes teaching mathematics a complex and wondrous thing.

Reference

NRICH 1997. 'Make 37', https://nrich.maths.org/make37.

19   Go Off-piste

Peter Ransom

Introduction

Motivating mathematical work tends to get harder as students get older, probably because mathematics starts to get more theoretical in preparation for GCSE. To counter this, go off-piste from time to time and deal with cases of practical mathematics that interest you. This helps to motivate students by providing them with reasons where mathematics is, or has been, useful. Throughout my career I have been interested in introducing historical cases where mathematics has been used. In this chapter I will give a few case studies using mathematics in an historical context, describing both the mathematics involved and the integration of other subjects. The national curriculum for mathematics states: 'Mathematics is a creative and highly inter-connected discipline that has been developed over centuries, providing the solution to some of history's most intriguing problems. It is essential to everyday life, critical to science, technology and engineering, and necessary for financial literacy and most forms of employment. A high-quality mathematics education therefore provides a foundation for understanding the world, the ability to reason mathematically, an appreciation of the beauty and power of mathematics, and a sense of enjoyment and curiosity about the subject.' I hope that these case studies provide some ideas as to how the above statement can be met.

Background

These case studies started life in the secondary school back in 2003 at the 60th anniversary of the Dambusters' raid (16 May 1943) when I finished working with a year 11 group before their GCSE exams. This final lesson with them was meant to bring together a number of mathematical topics in a different scenario. This went well and my daughter and I developed similar masterclasses (delivered in period costume) centred on the battle of Trafalgar (21 October 1805) to celebrate the 200th anniversary in 2005. Other cross-curricular work was developed on sundials (and delivered in the guise of John Blagrave, 1558?–1611, a gentleman of Reading), on mechanics (as Isaac Newton, 1643–1727), on fortifications (as Vauban, 1633–1707, the French engineer during the reign of Louis XIV), and on engineering (as Isambard

Kingdom Brunel, 1806–1859). These lessons in period costume were meant to inspire students by showing how mathematics was used at these periods in time and to help them realise that mathematics, science, technology, geography, history, English and music are all intertwined. It was intended that the mathematics covered should develop what they knew already and that they should sometimes be working at the frontiers of their knowledge when it came to applying that mathematics in different situations.

Sundials

The STEM subjects – science, technology, engineering and mathematics – are essential to the future economic growth of the UK and interest in those subjects begins when children start school. Working with sundials, observing how the shadow of the gnomon (the part of a sundial that casts a shadow) changes throughout the day, is perhaps the first evidence we have of the fact that the Earth rotates on its axis. Early humans would have noticed that the shadows of trees changed both length and direction during the day and that this varied throughout the year.

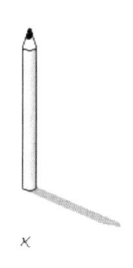

A preliminary activity to making a sundial is to make a pointer (a pencil stuck on a piece of Blu-tack, a bent piece of card etc.), and place it on a piece of paper in a sunny position. Now mark where the tip of the shadow lies. Students should put a mark where you think the tip of the shadow will be an hour later. Most students are quite surprised when they see how far the shadow has rotated. You can make it into a competition by seeing who is closest at the end of the period.

Use this idea to find due North, as follows:

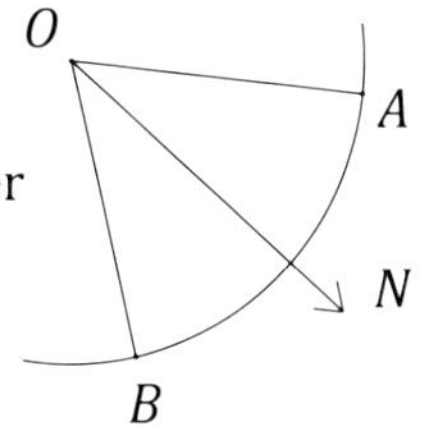

1. At some point in the morning (a couple of hours before noon) mark the tip of the shadow on the paper at A, say.

2. Draw an arc, centre O at the base of the pointer, radius OA.

3. Later in the day mark where the shadow next meets this arc. Call this point B.

4. Use the angle bisector construction to draw the line ON which bisects the angle AOB and then the line ON points to the North.

There is much else that can be done, such as constructing a sundial and calculating the angles using trigonometry. The Bowland Maths initiative contains a wealth of mathematical activity involving sundials, which I developed in conjunction with a group of teachers. You can have a look at the sundials case study at https://bowlandmaths.org.uk.

Students enjoy playing games and I have used the set of 30 sundial 'cards' (three of which are shown here) from the Bowland case study to encourage their number skills. They share the cards out and then use them to compare categories as in the Top Trumps game. For example, the student with the higher rarity (or largest dimension etc.) wins the card from the student with the lower rarity (or smaller dimension).

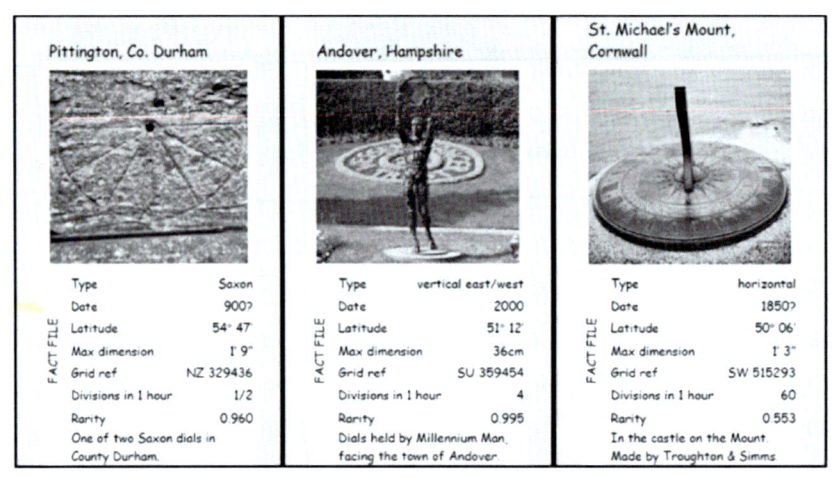

	Pittington, Co. Durham	Andover, Hampshire	St. Michael's Mount, Cornwall
Type	Saxon	vertical east/west	horizontal
Date	900?	2000	1850?
Latitude	54° 47'	51° 12'	50° 06'
Max dimension	1' 9"	36cm	1' 3"
Grid ref	NZ 329436	SU 359454	SW 515293
Divisions in 1 hour	1/2	4	60
Rarity	0.960	0.995	0.553
	One of two Saxon dials in County Durham.	Dials held by Millennium Man, facing the town of Andover.	In the castle on the Mount. Made by Troughton & Simms.

Older students can go on to make a sundial from an empty clear plastic bottle (good recycling!) and lots of information on sundials can be found at the website of the British Sundial Society (www.sundialsoc.org.uk/). How often do your students take home something mathematical that they have made in the classroom? Why not let them take home a sundial?

Vauban and his fortifications

French fortification is an incredibly rich area for geometry. Marshal Sébastian le Prestre de Vauban, or Vauban as he is better known, spent years under Louis XIV, the Sun King, fortifying towns and cities in France. I was fortunate to acquire an old mathematics text (Du Chatelard, 1749) that contained a treatise on fortification and the plates intrigued me so much I developed a series of lessons based on them.

The plate with which I start the work is shown on the next page. I use this plate with students asking them about the symmetries of the shapes and how they would draw them accurately. I give them the following information about the plate. It refers to campaign forts which are used when armies are on the move.

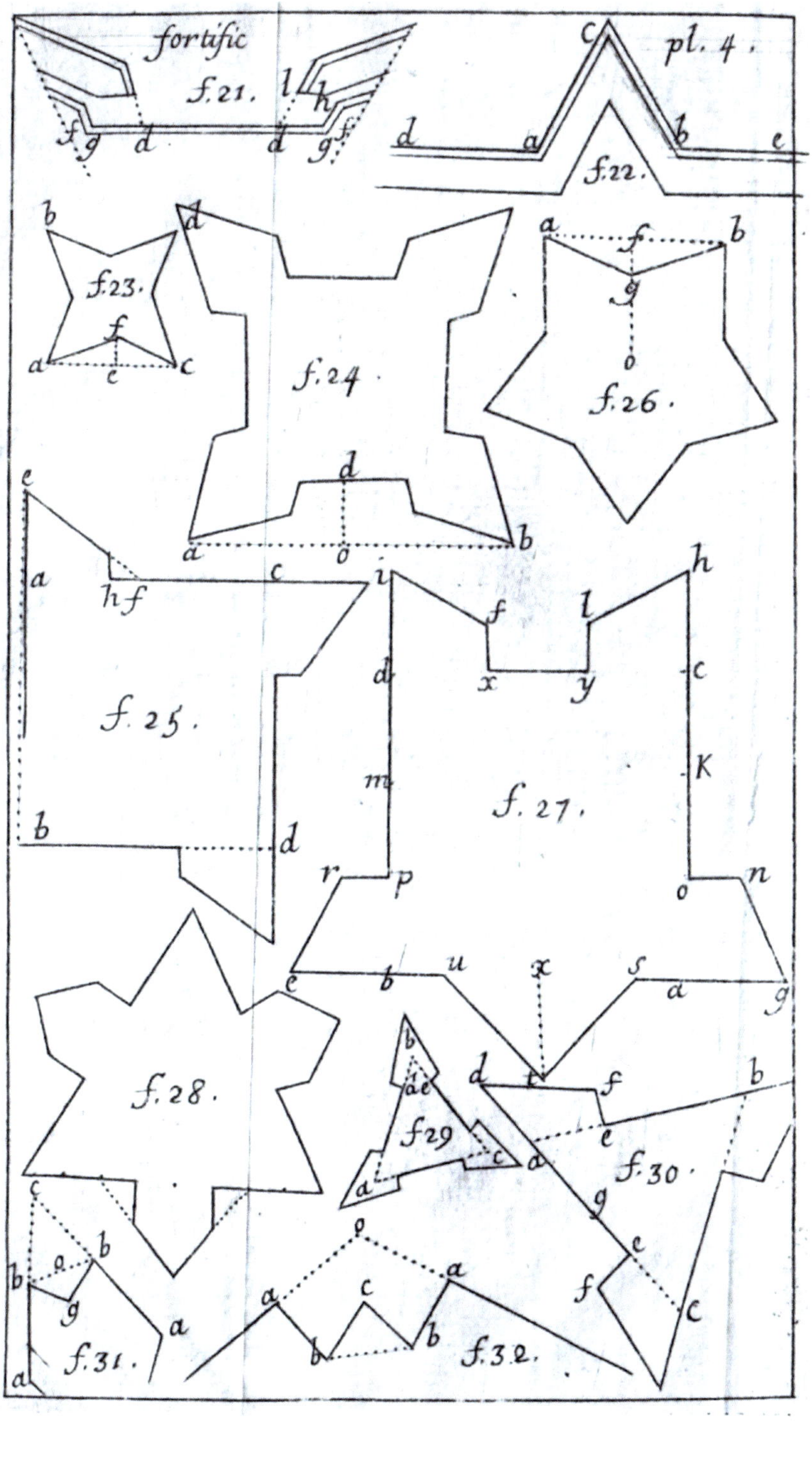
fortific
pl. 4.
f.21.
f.22.
f.23.
f.24.
f.25.
f.26.
f.27.
f.28.
f29
f.30.
f.31.
f.32.

I ask students to describe the symmetries of shapes f.23 to f.30. There is a short discussion about f.25 and we realise that the printer has probably deliberately omitted part of the fort at corner b to get the whole plate onto one page. The construction marks on one side of f.28 allow students to see how each side is divided to obtain the bastions. Students are asked to describe how this fort is constructed since mathematical communication is important. I ask students to draw the 'four star' fort of f.23, given that *ac* is 12 centimetres, and *ef* is 2 centimetres (Vauban's work states that the indent should be 1/6 of the side length), then to calculate both the perimeter and the area of the fort.

Trafalgar

This section was written up in *Mathematics in School* 39, 3 (May 2010), so all I list here are a few comments from students' feedback sheets, when they were asked 'What have you learnt?' The spelling remains the way it was written.

> *I learnt that when a cannonball is dropped from a certain height it can go 5 metres per second.*
>
> *It was brilliant, not only did I learn about (maths) But I learned some history too. If I could I would do it again.*
>
> *At school I find history dull but this was really interesting.*
>
> *I learnt about Trafalgar and the battle and ratio.*
>
> *I learned so much that I can't write it all down.*
>
> *I learnt a lot about Trafalgar and maths that I didn't know.*
>
> *I learnt about the battle of Trafalgar and how you can use Mathematics. If you put your brain to you can always do it.*

This work has proved popular with all learners. This is because they see a purpose to the mathematics – it is not just number crunching or pattern spotting for its own end. Linking it in to an historical event holds learners' attention. Working in small groups encourages learners to communicate with others. Students summarised their work in posters which allowed the others in the class to see what the different groups had done and how they could improve their work.

The development of this work has involved much risk taking, but that is all part of the enjoyment of teaching mathematics! Activities such as this and the visual and kinaesthetic links help consolidate learners' mathematical knowledge. The time invested in the use of these materials did in fact, saved time elsewhere.

Dambusters

This section was written up in 'The maths busters' (Ransom, 2004). One aspect of the work deals with measuring angles and distances: they are given a map and a description of the route taken by the Lancaster bombers and have to plot that route on the map, then describe the route in terms of the bearings and distances travelled.

Summary

One question I often get asked is how do I manage to get through the scheme of work when I sometimes spend two or three weeks on a topic. This is where flexibility comes into play. There is not enough time to do something on top of a scheme of work, so the cross-curricular work must include some of the topics in that scheme, though not necessarily from textbooks that might have otherwise been used.

There is much benefit in planning cross-curricular work with a colleague. Not only does it share the workload, but many useful ideas arise when there are two or more people thinking about what the work could encompass and how to bring in a range of subjects.

For cross-curricular work to have an impact in the classroom it must be of some significant interest to the teacher who delivers it. Then their enthusiasm for the topic helps empower the students in the classroom and they realise why they are learning certain skills and how these have been applied in the past. If the work has local interest, then relatives of the students are sometimes keen to get involved and can be a great source of information and artefacts relevant to the work. In particular cross-curricular work shows students the interconnectedness of subjects and that mathematics plays a major part in our lives.

References

Bekken, O. B., Mosvold, M. (eds.) 2003. *Study the Masters*, NCM, Göteborg.
DfE 2013. *Mathematics Programmes of Study: Key Stage 3*.
Dixon, L., Brown, M. and Gibson, O. 1984. *Children Learning Mathematics*, Oxford: Cassell.
Du Chatelard, P. 1749. *Recueil de Traités de Mathématique ... Tome Quatriéme*. Toulon: Mallard.
Fauvel, J., van Maanen, J. (eds.) 2000. *History in Mathematics Education: The ICMI Study*, Dordrecht-Boston-London: Kluwer.
Hambly, M. 1982. *Drawing Instruments: Their History, Purpose and Use*

for Architectural Drawings, London: RIBA Drawings Collection.

LePrestre de Vauban, S. 1968. *A Manual of Siegecraft and Fortification*, translated by Rothrock, G.A., Ann Arbor: University of Michigan Press.

Ransom, P. 2004. 'The maths busters – the geometry of the Dam Busters', *Mathematics in School* 33 (2), 22-24.

20   *Develop Your Questioning Skills*

Tom Button

When I started teaching there seemed to be a huge number of things that I needed to focus my attention on when planning lessons: whether I understood the maths, whether I had too much/too little content for the lesson, whether the lesson was too 'easy' or too 'hard' ... the list felt endless. However, as my teaching developed, I increasingly noticed that one thing had a disproportionate effect on the quality of my lessons: the questioning I was using when teaching the whole class. Two or three well-planned questions had the potential to help students develop their mathematical thinking far more effectively than any number of worked examples.

My own skills in questioning were enhanced when I was introduced to the wonderful book 'Thinkers' by Chris Bills, Liz Bills, Anne Watson and John Mason. If you've not read this, I'd highly recommend getting a copy. It's the definitive guide to using open questioning in maths. I think that, once you've understood the maths that you're going to be teaching, then effective questioning is the single most important skill for a maths teacher to develop. Good questioning is also often the essential ingredient to ensure another chosen teaching strategy will work, whether that is using mini whiteboards, using worked examples, or using technology.

Effective questioning when using technology

In 'Can I tell you one thing' Ben Sparks talked about the importance of making maths dynamic. Many free mathematics software packages, such as Desmos, GeoGebra and Autograph, feature the functionality of making a graph dynamic: you can enter equations like $y = mx + c$ then see the effect of changing m or c as movement on the graph. This can be a very powerful tool for helping students think about variation but, without directing the students' attention to *how* the graphs are changing, much of the potential of this will be lost. This is where well-planned questioning comes in: there are some straightforward strategies you can use when planning your questions about dynamic graphs that will help students make connections.

Over many years of working with teachers on helping them make the most of technology I have observed the power of three simple question types that can be used with any dynamic file. They are:

- How does it change … and why?
- For what value of … will …?
- How will it change?

I'll suggest some ideas for how you would use each type of these questions separately.

How does it change … and why?

To see the first question in action I'd suggest exploring the intersection of a line and a parabola. You can try this yourself now using your graphing tool of choice. In your graphing tool:

1. Plot the graph of $y = x^2$.
2. Plot the graph of $y = mx + c$. Note that most dynamic graphing tools should automatically set m and c to constants that can be varied with a slider or constant controller.
3. Set the value of m to 2 so that your line is $y = 2x + c$.
4. Add the point $(0, c)$. This makes it easier to see how the line moves.

Now you can change the value of c and see how the graph moves. You can even animate it so that it changes continually. I would like to use this to help the students think more deeply about the point of intersection of a line and a parabola, so the version of the "How does it change … and why?" question I will ask them is:

> I would like you to focus on the number of points of intersection. When I change the value of c how does the number of points of intersection change … and why?

The *how* part of this question is the most accessible. For the responses to this I would expect students to be able to see that the number of points of intersection can be 0, 1 or 2.

For the *why?* I would like the students to be able to link this to solving the equations simultaneously. Setting $x^2 = 2x + c$ allows them to think about the number of real solutions to the equation $x^2 - 2x - c = 0$ which can be related to the situations where there are 0, 1 or 2 points of intersection.

This question type is the best one to try first. I'd suggest exploring this for a few different examples with students before trying the other two question types.

For what value of … will …?

Once students are familiar with dynamic graphs you can try them with the second question type. For this you need to set up a situation with a dynamic graph but, instead of moving it, you will ask them to predict a value for which a certain situation will occur.

As an example of this prediction task, I'm going to stick with the graph of $y = x^2$ and $y = mx + c$ that I used above. I'm now going to change the slider for m so that it has the value of 4 and ask the students to predict the value of c where the line and curve have a single (or repeated) point of intersection. The question will be:

> I've set m as 4. For what value of c will the line and parabola have a single (or repeated) point of intersection?

This is a less open question but when used as a follow-on from the earlier type of question it allows for students to apply their understanding and obtain an answer. It also provides an opportunity for students to test their mathematical understanding of a situation, even if this is not fully developed. It is particularly powerful when combined with strategies such as 'think, pair, share' where students can discuss the method that they've used to find their answer as well as just stating it. You can often get different methods from different students and comparing these can be very enlightening.

One of the great things about this type of question is that the answer can then be verified dynamically by changing the value of c.

How will it change?

The third type of question is one that I'd leave until students were familiar with both of the other question types. Students need to be comfortable with the idea of a graph moving before you try this with them. The idea here is for you to set up a situation with a dynamic graph in it then get them to predict how it will change when you vary one of the values in the equation of the curve.

I'll stick with the same example of $y = x^2$ and $y = mx + c$ that I used above. In this case though I'm going to set c to have a value of 2 and change m. I would like the students to think about how the line will move and then use this to predict the number of points of intersection. The question is:

I've set c as 2 and I'm going to vary m. How many points of intersection can there be between the line and the parabola?

Once the students have committed to an answer (and not before!) you can then vary m and see if they are correct. As with the other types it's important to get students to explain their reasoning here.

A stepping stone to generalising and greater rigour

The idea behind all these question types is to encourage students to reason mathematically. The use of dynamic graphing provides a medium for this that emphasises the link between graphs and algebra but is also visually engaging. It is important to encourage students to express their understanding and reasons in their own words. They will often explain their understanding of mathematical relationships in quite an informal way initially. By using their answers, you can help them reformulate their ideas in a more formal and rigorous way. This is especially helpful in A level mathematics where one of the overarching themes is *Mathematical argument, language and proof.*

My experience of using these types of questions extensively is that they generate a far deeper discussion than closed questions. In doing so they allow for deeper misconceptions to be addressed that would otherwise be missed. So, if I could tell you one thing, it would be to develop your questioning skills (once you've understood the maths)!

References

Bills, C., Bills, L., & Watson, A. 2004. *Thinkers: A collection of activities to provoke mathematical thinking* (J. Mason, Ed.). Association of Teachers of Mathematics.
Southall, E., (ed.) 2022. *If I Could Tell You One Thing*, Mathematical Association.

21   *Take Time to Explain Why*

David Miles

Every now and again, a student will say, "I enjoy your lessons because you always make sure we understand why things are the way they are." Delivering explanations that are thorough, interesting and accessible are at the heart of my teaching and it is gratifying when this is acknowledged.

Clearly, every mathematics teacher wants their students to feel like this, so why do some colleagues choose not to embrace obvious opportunities for illumination? Over the years, I have witnessed the following by way of justification:

- "With a packed curriculum, I can't afford to waste valuable lesson time on unnecessary exposition."
- "It isn't fair to expect a non-specialist to go into so much depth."
- "Proof is fine for able students but most of my group wouldn't follow the reasoning and it would be counterproductive to damage their confidence this close to the exams."
- "During my training, I was advised not to get bogged down in lengthy explanations as they can make children lose focus and misbehave."

Whilst I understand these arguments, I don't accept them. Sufficient curriculum time can always be found, steps can be taken to improve subject knowledge and we should not allow the looming prospect of exams to have a negative impact on teaching and learning. Most importantly, young people should never, ever, be underestimated.

To illustrate my point, here are three explanations that, with appropriate scaffolding, can be appreciated by a wide range of students and may help them to build connections that increase their engagement in our awesome subject.

(1) Area of a trapezium

Assumed knowledge

- Area of a triangle
- Distributive property of multiplication

Explanation

Consider a trapezium with parallel sides of length a and b and vertical height h.

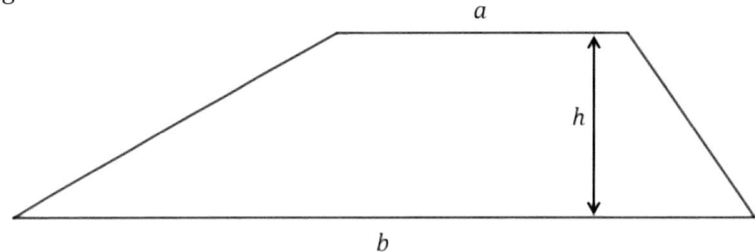

Divide the trapezium into four right-angled triangles as shown.

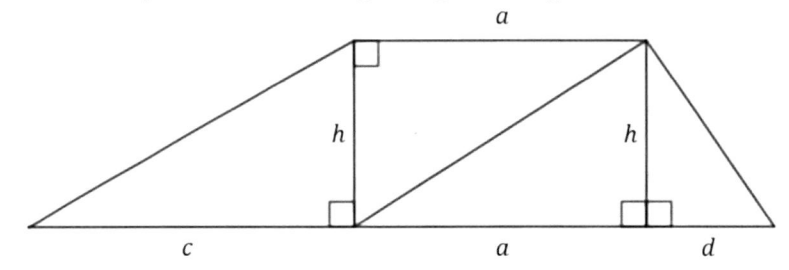

$$\text{Area of the trapezium} = \frac{1}{2}ha + \frac{1}{2}hc + \frac{1}{2}ha + \frac{1}{2}hd$$
$$= \frac{1}{2}h(a + c + a + d)$$
$$= \frac{1}{2}h(a + b).$$

Extension

Derive the constant acceleration equation $s = \frac{1}{2}(u + v)t$.

(2) Quadratic formula

No matter which GCSE mathematics set I am teaching, I always prove the quadratic formula. Doing so allows students to glimpse the power of algebra and it can provide some with a real lightbulb moment. I demonstrate how to apply the formula first and allow them plenty of time to become familiar with it. Someone will inevitably ask, "How did anyone ever come up with this?" They are often pleasantly surprised when they discover they already know enough mathematics to understand how the formula is derived.

Assumed knowledge
- Rearranging and manipulating equations
- Factorising a quadratic

Explanation

Take the general quadratic equation: $\quad ax^2 + bx + c = 0$

Multiply each term by $4a$: $\quad 4a^2x^2 + 4abx + 4ac = 0$

Add $b^2 - 4ac$ to each side: $\quad 4a^2x^2 + 4abx + b^2 = b^2 - 4ac$

Factorise the left-hand side: $\quad (2ax + b)(2ax + b) = b^2 - 4ac$

Take the square root of each side: $\quad 2ax + b = \pm\sqrt{b^2 - 4ac}$

Subtract b from each side: $\quad 2ax = -b \pm \sqrt{b^2 - 4ac}$

Divide both sides by $2a$: $\quad x = \dfrac{-b \pm \sqrt{b^2 - 4ac}}{2a}$

In the last step, we note that $2a$ is necessarily non-zero.

Extension

Develop a more direct proof by completing the square.

(3) The Pythagorean identity

Assumed knowledge
- Pythagoras' Theorem
- Definitions of the trigonometric ratios

Explanation

Consider this right-angled triangle.

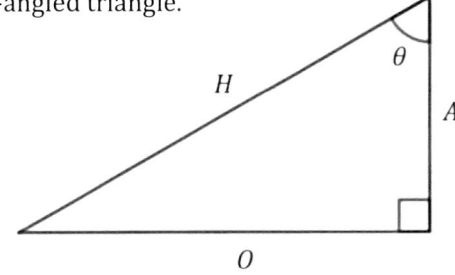

From Pythagoras' Theorem: $O^2 + A^2 = H^2$

Divide each term by H^2: $\dfrac{O^2}{H^2} + \dfrac{A^2}{H^2} = 1$

Now spot the trigonometric ratios: $\sin^2 \theta + \cos^2 \theta = 1.$

Extension

Explain why $\tan \theta = \dfrac{\sin \theta}{\cos \theta}$.

22 *Better Formative Assessment Review*

Darren Carter

The problem

Good assessment lies at the heart of a good mathematics education. We spend significant amounts of time planning assessments, preparing students for assessment, and marking them. However, in my experience, we often don't spend enough time on how to effectively review assessments. This seems like a missed opportunity.

As an early career teacher, I used to give very little thought to how to review an exam paper. I would simply stand at the front of the class and talk through the whole paper and the students would listen. As I gained more experience, I learned to ask better questions and provide better explanations but these lessons still felt ineffective to me. For an hour, I spoke and they listened.

I believe that the traditional approach of talking through the entire exam paper can leave some students feeling left out. For example, the student who scored 90% may have to sit and listen to questions on which they scored full marks, while waiting for a part of the paper where they need help. This can be frustrating for them, and it's important for teachers to find ways to engage and support all students, regardless of their excellent performance on an assessment.

It's natural to believe that students who have performed poorly on an assessment need clear explanations in order to improve. However, it's important to recognise that simply talking through an entire exam paper in a short period of time may not be the most effective way to help these students. If they were able to learn effectively in this manner, it's likely that they wouldn't have performed poorly in the first place.

Another issue with traditional exam review lessons is that they can promote passivity among students. Even when students are actively listening and copying down solutions, they may not be doing much mathematical thinking. This can be problematic, as it's important for students to engage with the material and develop their own understanding of it.

Finally, after an hour of working through an exam paper in a traditional review lesson, both the teacher and students may feel mentally exhausted. While this isn't necessarily a problem, it's worth considering whether this approach is effective enough to justify the time and effort involved. In some cases, taking an alternative approach may be more successful at helping students learn and retain the material. Teachers should carefully consider the goals of the lesson and the best ways to achieve them.

How I now review assessments

I have found that changing the approach to reviewing assessments can lead to better results and more enjoyment for both myself and my students. The approach that has worked well for me is to have a lesson following the assessment where students are asked to actively think through the questions again, with appropriate scaffolding and support to help them progress. This allows students to engage with the material more actively and can help them develop a deeper understanding.

After marking an assessment, I print out a copy and handwrite additional information onto the questions to help students have a second try. This can include things like partial solutions, probing questions, hints, or diagrams that provide scaffolding. By providing this extra support, students can understand the material better and learn from their mistakes. This approach has been effective for me and I've found that it can lead to more engaged and productive review lessons. By printing out these hints for students, they can focus on their own areas of development while I circulate the room providing support and guidance. This allows students to learn at their own pace and can help them develop a deeper understanding of the material.

What it looks like

Below are just a few of the questions I have annotated recently. The style and purpose of my scaffolding varies massively from question to question. The intent can be as simple as trying to draw students' attention to a specific part of the question I felt they missed, or provide more direct support like probing questions or diagrams. Overall, my goal is to help my students to improve their understanding and engagement with the questions at hand.

(1) In this question, students seem to either answer it correctly or leave it entirely blank. I decided to draw their attention immediately to the issues and even included a question to help structure their thoughts.

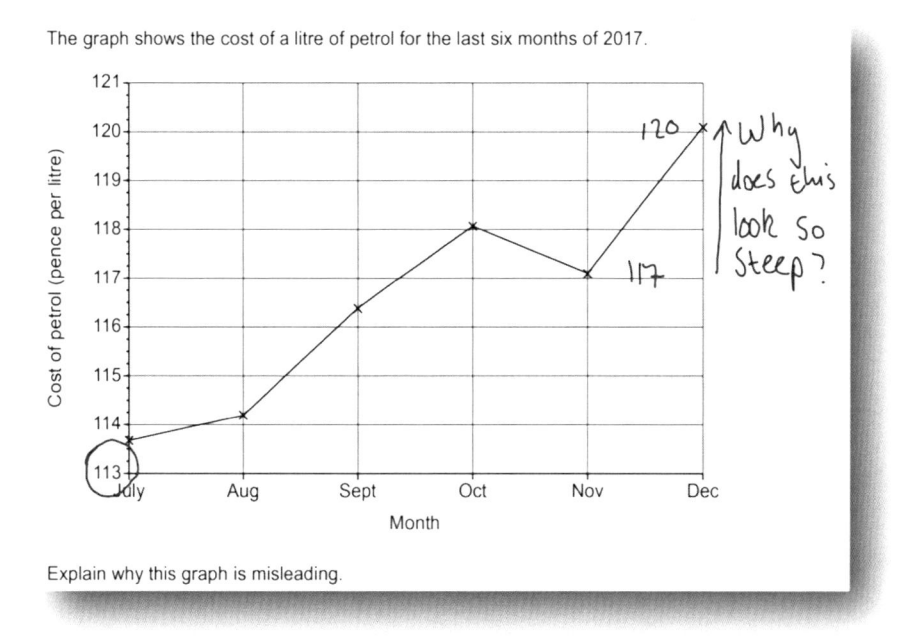

The graph shows the cost of a litre of petrol for the last six months of 2017.

[Handwritten annotations on graph: "Why does this look so steep?" pointing to the 120 value; 113 circled on July]

Explain why this graph is misleading.

(2) For routine questions, I often include familiar terminology or the start of a method in order to jog the students' memory on the topic:

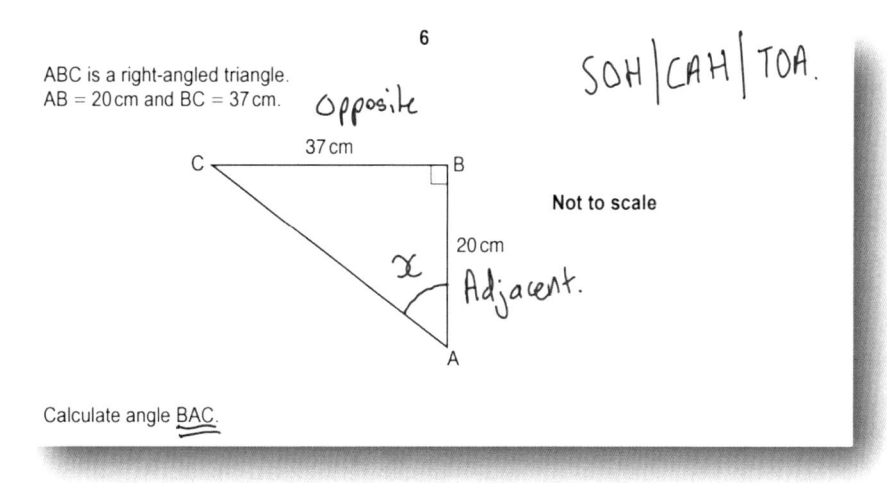

ABC is a right-angled triangle.
AB = 20 cm and BC = 37 cm.

[Handwritten annotations: "Opposite" labelling the 37 cm side, "Adjacent" labelling the 20 cm side, "SOH | CAH | TOA"]

Not to scale

Calculate angle BAC.

(3) When dealing with multi-part questions, I often use simple headings to help the students think about what they should calculate

113

rather than immediately telling them how to calculate. You can also see in this example the use of a diagram which would be familiar to my students, again to help them structure their thoughts.

> 6   A bag contains some counters.
>
> - There are 300 counters in the bag.
> - There are only red, white and blue counters in the bag.
> - The probability of picking a blue counter is $\frac{23}{50}$.
> - The ratio of red counters to white counters is 2 : 1.
>
> Calculate the number of red counters in the bag.
>
> *Number of blue counters?*
> *Number of red + white counters?*
>
> *Red : White*
> *☐ ☐ : ☐*

Practical tips

I always create these help sheets immediately after marking the assessment at the moment when I have the best understanding of where they went wrong, allowing me to provide the most succinct support possible. I either print out a copy for myself and handwrite onto it or load up a pdf on my tablet. This has never taken me very long to complete. I have often already thought about what I will be adding to the question during marking. It may even make sense to do this task alongside marking the paper, something I may try in the future.

While sharing help sheets between teachers has a place, it is preferable to make them yourself as it ensures that the support provided uses the same methods, diagrams and language that you have used in class. This makes it easier for students to understand and apply the information, as it is familiar to them.

After adding hints to each question, I then print the paper four pages per side, double-sided, on an A3 piece of paper. The students then share these sets in pairs. Now they have the hint sheet, they are given a whole hour to work through the assessment for a second time. Once they have got stuck in, I circle the room looking for opportunities to provide insight. I will also on occasion stop the whole lesson to go through a particular question in a more traditional manner if I feel it is required.

Final words

Why do I really enjoy this approach? It encourages a room full of mathematical discussion. Simply providing the students with their papers and some hints leads to them taking over; they are explaining to each other, they are thinking deeply, they are asking questions, they are getting stuck in.

I hope this has not come across as if I am dismissive of the impact that teacher exposition can provide, sometimes expert explanations are required. This approach facilitates rather than opposes exposition. As everyone is working, I will be dipping into conversations and adding any input I feel is necessary.

This is one of the many changes I have made to my practice over the years. I am sure there's many more to come. I think the key in teaching is to just give it a go, evaluate and adapt.

Reference (Source of questions)

AQA GCSE Mathematics Higher Tier Paper 3 (Summer 2019).

23   *Numeracy and the Importance of Maths Teachers*

Susan Okereke

I love teaching maths! I believe that, with good instruction, a 'can do' attitude and kindness, everyone can improve their maths skills and it feels great when this happens. There is a common misconception that maths is just about getting the correct answer but it is so much more than that. Maths is about seeing patterns, making connections and solving problems. It is about communication and often people need to think and work together to solve these problems. It can be an incredibly creative subject. Maths is also an emotional subject.

I completed a Masters in Teaching a few years ago and it highlighted the transformative nature of education, especially maths education. The potential opportunities that studying maths can give people today are huge, with the growing need for workers in STEM (Science, Technology, Engineering and Mathematics) industries and increased job roles that require quantitative data analysis skills.

In a more general sense, every student should leave school competent and confident in everyday maths, in the same way as being able to read and write, but sadly, this is not the case for many students in the UK. Although I love maths, I am fully aware that there are many people (young and old) across the country that have been traumatised by the subject and suffer from 'maths anxiety', which has resulted in the often unspoken issue of poor numeracy, particularly amongst adults, which impacts our students, our economy and society as a whole. Poor numeracy is a huge problem for the UK (as I explain below) and my aim for this chapter is to encourage thought, discussion and collaboration because I believe that maths teachers are key to solving this problem.

Numeracy

The charity National Numeracy, set up to improve numeracy skills in adults in the UK, defines 'numeracy' as "the ability to understand and use maths in daily life, at home, work or school". The UK needs a numerate population in order to build a strong economy and to compete globally and everyone needs to be numerate to maximise their life chances and to make a positive contribution to society. Sadly, research carried out by National Numeracy found that 49% of the working population of England have numeracy skills equivalent to primary school students (National Numeracy, 2012). This is alarming as recent studies (National Research and Development Centre, 2013) have shown that numeracy is a bigger indicator of disadvantage than literacy, with people with poor numeracy skills being more than twice as likely to be unemployed as those competent in numeracy. Also, pupils beginning secondary school with very low numeracy skills but good literacy skills have an exclusion rate twice that of pupils starting secondary school with good numeracy skills. Poor numeracy poses a financial cost to the individual and is estimated to cost the economy £20 billion a year (National Numeracy, 2012).

This issue of poor numeracy in the UK isn't helped by the widespread negative attitudes towards maths in the media and wider society. This quote sums it up well,

> 'It is culturally acceptable in the UK to be negative about maths, in a way that we don't talk about other life skills. We hear 'I can't do maths' so often it doesn't seem a strange thing to say' (Kowsun, 2008).

Although this picture might seem bleak, there is much that can be done to improve things and it starts by improving people's attitudes towards maths. A National Numeracy report (2013) stated:

> 'Negative attitudes, rather than a lack of innate talent, are at the root of our numeracy crisis. In order for people individually – and the country as a whole – to improve and in turn benefit from raised levels of numeracy, our attitudes have to change.'

I believe maths teachers are the key to succeeding on this mission because teaching is one of the few professions which has the ability to change students' minds, but in order to maximise this potential for change it is important for us to understand the process of learning.

The science of learning

Developments in the understanding of the brain and the science of learning highlights that there are three parts to the learning process (www.scienceoflearning-ebc.org/): Engagement, Building Knowledge and Consolidation.

I'm going to focus on the idea of engagement because there are many teaching resources on building knowledge and consolidation but I believe we don't talk enough about how to engage our students in the learning process.

It is important to understand that the learning process begins with engagement. The aim is to encourage an 'approach response' in the learner such as 'there's something happening that I want to engage with' and thus the brain is primed and ready to take in information. Priming the brain for learning can be done by using rewards. This doesn't have to be sweets or money. Rewards are strategies that encourage students to get 'turned on' to learning, like praise, a token acknowledging achievement, novelty, group activities, giving students a choice of different tasks. For different students, different rewards will be effective, so we teachers should employ many different strategies. This is basic learning theory – a reward is anything that reinforces a behaviour.

Alternatively, if students are feeling fear or anxiety an 'avoidance response' occurs and they are 'turned off' to learning and they want to move away from the learning situation. If that is the case it doesn't matter how well you've planned an activity, if the students are anxious or fearful, they cannot engage with the learning process. It is important to deal with anxious feelings before attempting to proceed with the learning activity.

Engagement can be reduced by teachers' and students' beliefs that there is a fixed limit to what a student can learn. Students' core beliefs about learning influence their response to challenge, so changing a student's attitude to maths is about getting them to believe that they can do the work. A positive mindset is essential for learning to take place. If a student believes they cannot, they will not. If they believe they can, there is a chance that they will; so it is important for us as teachers to share with them the belief that ability comes through effort rather than innate aptitude. We need to encourage our students to buy into the idea that struggle and perseverance are essential parts of learning and on the other side, once they've completed a task, the success feels fantastic. Teachers and parents can teach students to have this kind of 'growth mindset' by showing them we believe in their ability and by praising their effort and resilience.

This knowledge has transformed my teaching because I now spend time thinking about how to cultivate a learning environment that is engaging, welcoming and safe;
- an 'engaging' environment because we now know the learning process starts with engagement and different students will be

engaged by different things, so this involves getting to know my students and trying out different strategies.
- a 'welcoming' environment because students come to my class from a variety of places with a variety of mindsets, and it's important that they know that I believe in them and that I like them. You don't have to be their best friend but it's about letting them know that you appreciate them as individuals and that you understand the fact that they have a unique brain and unique needs.
- a 'safe' environment because challenge is part of the learning process, especially in a maths classroom, and it is important to create a space where learners understand that they will make mistakes and they will find tasks difficult, and that is okay and necessary for learning to take place. It's about cultivating a space where difficulty and challenge can arise and students feel safe to give things a try, while also encouraging each other. Failure is a part of the learning cycle, but too few students feel safe enough to tolerate this and understand that it is a part of growth.

I think it is important to take a moment to acknowledge the importance of students' parents or guardians in the learning process. Parents have the biggest impact on students' lives and the messages they get at home are the inner voices they bring to school, so if a student hears 'I can't do maths' at home, this will have an impact on their attitude to maths. In my view it is helpful for parents to be given consistent messages from teachers and schools such as:

- ✓ Please talk positively about maths, especially when a task is challenging
- ✓ Where possible, highlight maths in everyday life, like in cooking, in decorating, in the design of buildings, in the stars in the sky, etc.
- ✓ Please praise your child for their effort rather than talent. By praising their effort, you are supporting them to be lifelong learners.

I write this chapter fully aware of the challenges faced by the education system, including a national maths teacher shortage and many schools facing tighter budgets which is impacting the resources available to teachers but I write this chapter with hope. We teachers are still the masters of our classrooms and we have the power to welcome our students into a world of maths that is engaging, safe and fun, setting them up to go into the world as confident, competent and numerate adults. If I could tell you one thing, it would be to focus on engagement.

References

Dweck, Carol. 2008. 'Mindset and Math/Science Achievement', *Teaching & Leadership: Managing for Effective Teachers and Leaders*. www.growthmindsetmaths.com/uploads/2/3/7/7/23776169/mindset_and_math_science_achievement_-_nov_2013.pdf

Kowsun, J. 2008. 'This innumerate isle', *TES*, online at www.tes.com/magazine/archive/innumerate-isle

National Numeracy. 2012. 'A New Approach to Making the UK Numerate', online at www.nationalnumeracy.org.uk/sites/default/files/documents/nn124_essentials_numeracyreport_for_web.pdf

National Numeracy. 2013. 'Attitudes Towards Maths – Research and Approach Overview', online at www.nationalnumeracy.org.uk/sites/default/files/documents/attitudes_towards_maths/attitudes_towards_maths_-_updated_branding.pdf

National Research and Development Centre for Adult Literacy and Numeracy. 2013. 'The Impact of Poor Numeracy Skills on Adults', online at https://maths4us.files.wordpress.com/2013/08/nrdc_impacts-of-numeracy-review_june13-m4u.pdf

STEM Learning, *The Science of Learning* (NE709), online at www.stem.org.uk/cpd/ondemand/422179/science-learning

24   *Examiners are your Friends*

Graham Cumming

If I could tell you one thing, it would probably be a dozen things and it would really be directed to your students but perhaps, as their teacher, you could pass it on? It's that the examiners are your friends. It might not feel like it sometimes but more than anything they want you to do well in your examinations. So why do they set such difficult questions?

They shouldn't all be difficult – papers are designed so the questions get a bit harder the further you go through them. It's a bit like when you arrive for work; other than the emergency services, few people do their hardest work the moment they arrive. Mostly they say hello to any colleagues already in and go to make a coffee and have a bit of a chat. Think of the first few questions in a paper as the coffee machine – you need to know how it works, but that won't be the toughest part of your day (and if it is, congratulations on your cushy job!).

Early questions are designed to instil confidence; they are the ones examiners expect you to be able to do as if you were limbering up for a longer challenge and indeed, that's what you're doing.

A bit further in you're probably going to find the first of the questions set in context. This will be a problem that you might encounter in real life, such as working out your change on a purchase or the speed of a car pulling a caravan up a slope. Using maths in context is one of the principal skills expected by the curriculum; this is how problems come at you in real life. It's rare that anyone will give you the equations involved; in such circumstances, you have to work out what they will be and solve them.

Examiners are teachers, or they have been teachers; they have a good idea of what students should be able to understand. Not every context is one you are going to have experienced yourself, but they are generally ones you will be aware of (for example, filling up a car with petrol, going on holiday, renting a van). It doesn't always matter. What matters is that you can turn the information into something mathematical and solve the problem that way; it's why you learn it at all. Your school education is often about preparation for real life and for many of us, that life won't always be easy – give yourself the tools to navigate it as best you can.

Examiners meet to discuss exam questions and look at every single word, every single piece of punctuation. They make sure that sentences are short and unambiguous. They make sure that words used aren't technical (other than those which have a particular mathematical meaning which you're expected to know). If they can't do that, the question often gets thrown out and a rewrite is requested and they'll do that until all examiners are happy with it. So your first reaction to a question in context shouldn't be "I can't read this" – chances are you can, but there's still some thinking to be done to turn it into maths and that can be the bit that makes your brain hurt. Make sure you practise this skill so you can turn contexts into maths to be solved, even if you haven't experienced them; you might never have tiled a bathroom but you might have to one day.

Examiners work hard to make sure the information needed to answer each question is given, whether in words or diagrams or both, and all of it will be relevant. Honour the work of your examiners by making sure you always read the question – the whole question. Skip nothing.

As an aside, there are no reading ages for maths papers. No-one measures them and anyway, the algorithms used on long prose pieces don't work on short maths questions. The only people who refer to reading ages for maths exams are marketing departments and they should know better.

As mentioned before, papers are ramped in difficulty and you'll start to feel that as you progress through; like walking up a steep hill or staircase, questions will get tougher as you go. There can be up to half a million students sitting a maths paper and this is the way examiners can find out which ones get the top grades, middle grades and lower grades – but they will still do all they can to help you get the marks. All students will have had different experiences before they reach the exam; their schools, their teachers, their home life, whether or not they've put the work in. The playing field can be pretty uneven for 16 years, but it levels out for a brief hour or two when the exam is on.

The staples in the middle of a paper are pretty much the halfway point on an examination paper and, if you've been entered for the correct tier or option, this should be where you start to feel the pressure a bit – you're halfway up the hill. If it was a day at work, this might be when you stop for lunch but not this time – you'll need to power through. It's good to get some practice in at working uninterruptedly for an hour or two – no phones, no internet; do what you can to build up stamina.

There will be more questions in the second half about reasoning and problem-solving which require a little thought before putting pen to paper – what are they?

Reasoning questions will require you to offer a few thoughts about how you have arrived at your answer. They're not always popular – "why are we having to do English in a maths exam?" is a common complaint – but examiners will want to know you understand as they do. They won't insist on perfect grammar or spelling, so don't let that inhibit you – if you know a reason or an explanation, write it down when asked.

Problem-solving questions are what they say – a problem to be solved. You could say that if you can solve it, it's not a problem and there's some truth in that but otherwise, there will be a little working out to be done. Some questions are broken down into parts (a), (b), (c), etc which will often give you a direction to go in and you should always try part (a) because that may well be the key to the whole thing. Other questions will just give you the information needed and leave you to it – these questions do need some thought and, in a sense, you will need to put in your own parts (a), (b) and (c). Make sure you try to get a first mark on these questions, deduce some information that might lead you to gaining the second, third and further marks available.

If you've made it to the end of the paper, congratulations! Make sure you look at the back page as well – if there's no question there, it will say BLANK PAGE so you know there hasn't been a printing error. Examiners use the back cover sometimes to avoid having to add another four pages to the paper, which would equate to about 5000 reams of A3 with a weight of 5 tonnes to move around the country.

When examiners come to mark what you have done, they will still be hoping to give you marks where they can. Sometimes they can give marks for a method even if you have made an arithmetic slip, quite often they can allow exotic spellings. That said, do anything you can to show examiners what you are doing (for example, setting out your working neatly or writing down an occasional sentence of explanation). Remember that examiners have many questions and papers to mark over a short period over the summer – make their job as easy as you can for them!

The best exams are those where there are no tears and students come out reporting that "it wasn't so bad after all" or even "I think I did quite well there." Examiners try their best to write those exams and look forward to you doing as well as you can.

25 Using Whiteboards in the Mathematics Classroom

Rob Southern

Anyone who follows me on Twitter will know that I am always banging on about A3 whiteboards and what a powerful resource they are for teaching A level Maths. In my first job, I was fortunate to have a mentor who taught using whiteboards and I was always in awe of how purposeful his lessons were, how independent his students were and the wonderful collaborative environment that he fostered among the students in his class.

Mini-whiteboards are most commonly used as a check for understanding, often in a 'my-turn-your-turn' scenario. They provide immediate feedback on what the students can do. However, they have so much more potential than this, particularly in the A level classroom. They allow the students to explore different approaches, test ideas and assumptions and make predictions. If they make a mistake or their approach doesn't work, they can erase it – no problem.

The visibility of the whiteboards encourages collaboration. Students can support and challenge each other and share their ideas. It is vital that the room is set up to facilitate this. The students need to sit so that they can see as many other whiteboards as possible. No hiding at the back!

Initially, the students need to be encouraged to share their ideas. The best way to do this is to get them to compare answers. If they agree, have they approached the problem the same way? If they disagree, can they work together to resolve the issue? If their approaches are different, which was more efficient? Will this approach always be the most efficient for this type of problem? Hopefully, this will lead to the students being less dependent on the teacher. The colleague I mentioned would always say in response to the question, "Is this right?", "I've got twenty A level mathematicians in the room, so why should I work out the answer?"

Support and challenge can be provided subtly. If a student is struggling to get started, I can write a prompt on their board. If they have finished quickly, I can suggest an extension. The whiteboards mean that there is no hiding place. All students have to be actively involved in the lesson.

While some students find this scary and they worry that you are going to see them getting it wrong, it is the best way to provide immediate, targeted support. If this is done within the context of a supportive classroom environment where students are encouraged to help each other, everybody wins.

Crucially, the whiteboards can show us what we actually *need* to teach the students. When I use whiteboards in A level lessons, the idea is to try to establish what the students already know, so that I can teach them the bits that they don't. So rather than starting a lesson with a teacher-led example, I will put a problem on the board and let them have a go, prompting and filling in the gaps as necessary. I might intermittently add a line of working to the example on the board. This can serve as a check for the students who have got past this stage, or a prompt for those who haven't.

Now, I am not saying that it is never appropriate to model something from scratch. This is an absolutely crucial part of introducing new content. However, once you have modelled the general approach, I believe that every question you subsequently do with an A level class should be moving them forward in their learning. An example of this would be when teaching integration by substitution. I would start by carefully modelling an example such as the one below:

Use the substitution $u = 2x + 1$ to find

$$\int x(2x+1)^6 dx.$$

However, I would very quickly be posing a problem like this for the students to attempt on their whiteboards:

Use the substitution $u = \sin\theta$ to find

$$\int \frac{1}{(1-x^2)^{\frac{3}{2}}} dx.$$

Now, I am expecting the students to get stuck on this one and I am pretty sure I know where this will happen – moving from $(1-x^2)^{\frac{3}{2}}$ to $(\cos\theta)^3$. However, the students won't know where they are going to get stuck until it happens. The process of getting unstuck, perhaps with a prompt from me or a peer, will teach them something important about these integrals in particular. It is also important for students to experience being stuck more generally so that this is not something that frightens them. Different students will need different prompts to become unstuck.

If they engage with this process, they can produce bespoke notes that identify the bits of each concept that *they* found tricky.

An extension for students who do not get stuck might be to try to think of a different substitution or to construct a similar question of their own. This approach is preferable to saying "now we're going to do one with a trig identity" and modelling it at the front because, rather than being told what the key points are, the students have experienced them for themselves by hitting those roadblocks.

The whiteboards are brilliant for pointing out common mistakes to the students. One example would be a question like this:

Divide $2x^3 + x + 7$ by $x + 2$.

The common error here is for students to get in a pickle because the x^2 term is "missing". If you give this problem to the students, some of them will spot it and others won't. The ones who don't will learn more from having made this mistake than from being told in advance to avoid it. This approach prevents the students from being passive because they are continually being asked to engage with new content.

Another good opportunity to do this is with partial fractions. When first presented with a fraction with a repeated linear term in the denominator, such as this one,

$$\frac{x^2 + 8x + 9}{(x+1)(x+2)^2}$$

most students will do this:

$$\frac{x^2 + 8x + 9}{(x+1)(x+2)^2} \equiv \frac{A}{x+1} + \frac{B}{x+2} + \frac{C}{x+2}.$$

Allowing them to get it wrong and discussing why it doesn't work is so much more powerful than pre-warning them.

The same approach will work with improper algebraic fractions. With this one,

$$\frac{3x^2 - 3x - 2}{(x-1)(x-2)}$$

students will almost certainly do this:

$$\frac{3x^2 - 3x - 2}{(x-1)(x-2)} \equiv \frac{A}{x-1} + \frac{B}{x-2}.$$

We discuss why that doesn't work, fix it and move on.

The whiteboards also allow students to explore and compare different approaches. Sometimes they may surprise you with what they come up with. This question on vectors is taken from the Pearson Edexcel textbook (Smith et al, 2017).

> In the triangle ABC, $\overrightarrow{AB} = 3\mathbf{i} - 2\mathbf{j}$ and $\overrightarrow{AC} = \mathbf{i} - 5\mathbf{j}$.
> Find the exact size of $\angle BAC$ in degrees.

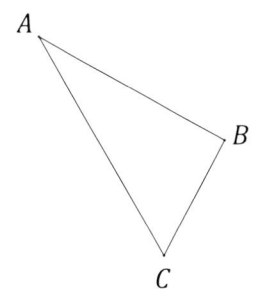

When I posed this problem to a class, the most popular approach was to use the Cosine Rule, which first requires you to find the vector $\overrightarrow{BC}$. In finding this vector, some students realised that $\overrightarrow{AB}$ and $\overrightarrow{BC}$ were perpendicular and equal in length and that this allowed them to conclude that it was a right-angled isosceles triangle without needing to use the Cosine Rule. One student did this:

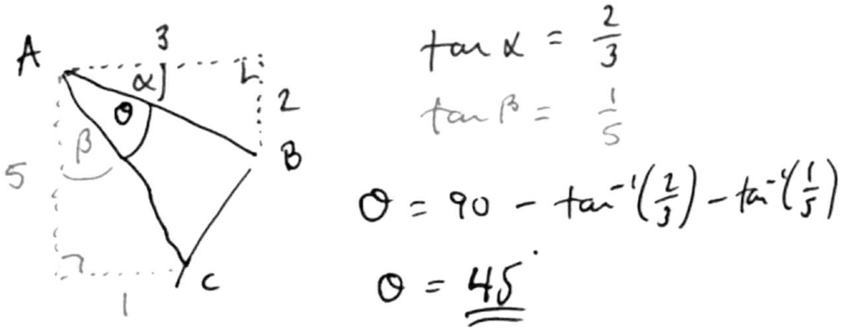

$$\tan \alpha = \frac{2}{3}$$
$$\tan \beta = \frac{1}{5}$$
$$\theta = 90 - \tan^{-1}\left(\frac{2}{3}\right) - \tan^{-1}\left(\frac{1}{5}\right)$$
$$\theta = 45°$$

This is an approach that I would never have considered. By using the whiteboards and encouraging the students to try different ideas, a novel approach emerged that enriched the experience of everyone in the room, including me.

The only solution given in the textbook is the Cosine Rule one. By sticking to a modelling approach, we can potentially restrict the versatility of our students. Also, the students who see things slightly differently feel that their thought process is validated by having the opportunity to share it with their peers.

Giving the students the freedom to surprise you can be scary. If you are going to teach this way, there are three words that you have to be comfortable saying to students:

"I don't know."

I used to be terrified of this happening to me in the classroom. I thought that the students would think I was a fraud and wasn't up to the job. However, throughout my career, some of the biggest improvements I have made to my practice have come as a result of a student asking a question that I couldn't immediately answer. I now say to students, you are working on *advanced* mathematics. How would you feel if every time you asked me a question, I immediately knew the answer as if it were trivial? You should be aspiring to ask me questions that I don't know the answer to and you should see this as a strength in you, not a weakness in me.

In summary, use whiteboards, let them get stuck, find out the bits they know and teach them the bits they don't. And enjoy the moments when someone suggests something that surprises you.

Reference

Smith, H et al. 2017. *Pearson Edexcel AS and A level Mathematics – Pure Mathematics Year 1/AS*, Pearson Education Limited, p. 245.

26 *Good Enough*

Mel Muldowney

Dear Reader,

So whilst reading the amazing contributions to this book, I suspect you've experienced emotions on a spectrum from, "Yeah, yeah, yeah, I do that" to "Wow! I can't believe I hadn't thought of that ... can't wait to try it out" (hopefully more of the latter than the former).

All new (to you or me) ideas can be brilliant; without ideas, progress is not made, and change does not happen, and it is important that as teachers we are innovative, creative and open to change. That said, it is equally important to retain what works whilst discarding or improving things that can be made to work in a better or more efficient way. Now let us be honest here, were you to try to introduce every new thing that you come across it would be chaos; you would never get to the point of reviewing what did and didn't work and you'd end up continually doing "all the things". In the long run this would be unsustainable, and we would lose someone valuable from the profession – yes, I mean you!

We all know that schools are hectic places. No two days are ever the same, in fact no two hours are ever the same. We all know how quickly the atmosphere can change within a classroom by the admission into the room of the tiniest insect or how the climate within the building can be affected by the smallest flutter of a snowflake outside the windows. Combine the ever-changing environment with all the usual things that happen routinely (learning walks, data drops, assessments, marking, CPD, reports, parents' evening, performances etc) and school life is just relentless.

Now don't get me wrong, I love teaching but unless you are careful you will find yourself living your life in blocks of six to eight weeks. Trust me when I tell you that you will find yourself saying, "Oh, I'll do it in the next holidays", when faced with anything that is outside of school life. This could include seeing friends or just the day-to-day life admin stuff, such as getting your car serviced or the grey hairs needing a touch up at the hairdressers. Even banking a cheque can prove a challenge as trying to get to a bank when they are open after a school day is nigh on impossible. There will be a point where living with a less than perfect house becomes the norm; a point where you consider your house is a mess (okay, I know that is relative, so I suppose I mean it is not as tidy as you would like) and you may find that there will be piles of textbooks "here, there and every-flipping-where" (my husband's words!). There will be a point where you constantly have a mountain of clothes waiting to be ironed or put away and at its most extreme you may even find yourself having sleepless nights and a constant feeling of dread about an impending visit from your mother-in-law. The nightmare where she is tutting as she looks behind your sofa checking to see if you've cleaned your skirting boards recently (if ever) will haunt you forever.

This cannot be right. There must be a better way; a way that means being able to give one hundred percent to our jobs but also have a life too.

In an age where social media is so prevalent, it would be easy to think that everyone else has this nailed down. We are bombarded with images of impeccably turned out families, having fun times together whilst holding down jobs that they are excelling at and having time for hobbies too. The social media view of the teaching profession, at times, suggests that there are thousands of teachers who have incredible classes all the time, making the most amazing progress in their neat and tidy classrooms where they are on top of everything that comes their way.

Despite knowing that what you see on Twitter or Instagram – other social media platforms are available – is only half the story. People share what they want you to see; it is hard not to make comparisons. However, I can assure you that almost everyone will go through a period where we don't feel as "good/smart/thin/fit/talented" as the people and teachers we are around or interact with. Okay, so maybe not those exact words but realising that this feeling of doubt happens to most people is the starting point of getting the balance right.

"Comparison is the thief of joy", is a quote attributed to US President Theodore Roosevelt and it suggests that the very act of doing so can risk sapping our pleasure in life, even eroding our happiness bit by bit, over time. Making comparisons with an unachievable ideal can make us less productive because we are too busy comparing ourselves to others instead of focusing on our own work. It can also create a sense of frustration, which can lead to us giving up altogether (such as not applying for your dream job because you do not think you would even get an interview). So, not only does the comparison make us feel bad, but it can also stop us from being productive and in the short term this will impact on us in the classroom too.

Many teachers are essentially perfectionists which, in the main, is a healthy form of perfectionism whereby they strive and thrive to achieve the goals they have set but at the same time, they will understand if all things are not accomplished.

We all have bad days and striving for perfection alongside feeling that you're just not good enough really is a perfect storm. After a bad day, effectively, you end up comparing your worst to everyone else's best that you are seeing around you or on social media which could lead to you feeling less than the perfect teacher. Don't try to run after a mirage, it is just not helpful – every time you think you are almost there, it will end up fading as you step closer to it. The same can be said when we are chasing our perception of "perfection" where we continue to work even harder, trying every new idea, ignoring our own health, stopping seeing friends etc but just like a mirage, it will end up fading away because "perfection" is just a misconception.

Striving for an unachievable sense of perfectionism has been found to be linked with various mental health conditions and other lifestyle problems. If you think this is affecting you, google "maladaptive perfectionism" and check out the symptoms. Several studies have been conducted and all of them have hinted at a negative correlation between maladaptive perfectionism and the overall well-being of a person. There are some simple strategies you can use to help you cope and discover a new version of "perfection". My fear is that ultimately we could lose you from the profession and the potential life-changing impact that you will have on students' lives as a maths teacher will be lost and we don't want that.

So, if I could only tell you one thing it would be that "Good enough, REALLY is good enough".

This is not an admission of defeat. Some people consider that starting out accepting anything less than perfection, anything less than the best is a form of failure … it really is not the case. It really is not accepting second best. "Good enough" does not mean that you lower your standards because having high standards is not the same thing as demanding perfection.

Teaching is one of those jobs, that if you wanted to, you could always find something to do, whether it is tinkering with a lesson or sorting resources or even marking. It is not like some other jobs, where once something is finished you move on. To thrive and not just survive in teaching we need to appreciate and accept the fact that it is okay if not everything is achieved with the perfectionism that was originally thought achievable. We need to stop looking for things that are not "just so" or introducing new ideas because that is what others have done and it ticked some box without firm evidence that there was a positive impact on outcomes for students. Making changes or introducing new ideas should be done with the students in mind, first and foremost, and if it "ticks a box" then it is a bonus rather than doing something *because* it ticks a box.

So, I leave you with my mantra: "Good enough REALLY … is good enough".

Remember, you are perfect as you are!

… oh … one more thing … always have emergency chocolate in a drawer at work!

Contributors

David Miles @Mathematical_A [Editor]

David is an Assistant Headteacher at Sir John Leman High School, Suffolk. He has taught secondary mathematics for twenty years, authored national and international assessments and acted as an A level examiner. David has produced A level teaching resources for The Mathematical Association and is currently Chair of the MA's Teaching Committee.

Rebecca Atherfold @becatherfold

Rebecca has recently joined MEI as a Maths Education Support Specialist in Further Education. She has taught in primary and secondary schools and FE colleges. Rebecca has a particular interest in giving students the space to explore maths: increasing their confidence and ultimately their enjoyment of the subject.

Amanda Austin @draustinmaths

Amanda is a secondary maths teacher and Key Stage 4 Coordinator with over ten years of teaching experience. She enjoys creating maths resources and using them in the classroom, sharing her resources through the website www.draustinmaths.com.

Tom Bennison @DrBennison

Tom is a head of Sixth Form in Derbyshire but his specialism is mathematics. Following a PhD in Computational Applied Mathematics, Tom trained to be a maths teacher and is passionate about increasing the numbers studying Post-16 maths and then going on to apply to study the subject at university. He is the author of the A level maths knowledge quiz book for John Catt and an editor for the Tarquin *A level Mathematics* series. He is also a Post-16 lead for the East Midlands West Maths Hub.

Tom Button @MathsTechnology

Tom is the Mathematics Technology Specialist for MEI. He is passionate about the potential of technology to deepen students' understanding of maths and has both designed resources and delivered many teacher courses on this. He also developed MEI's Further Pure with Technology unit and MEI's Data Science courses for students.

Contributors

Darren Carter @MrCarterMaths

Darren is the creator of www.mrcartermaths.com and the KS5 leader at Ridgewood School, Doncaster. His website is used in over 1000 schools across the world at KS2 to KS5. He has spoken at conferences, delivered CPD and most importantly loves the classroom.

Graham Cumming @MathsNot

Graham worked in a variety of roles at the Pearson/Edexcel examination board on mathematics qualifications at all levels, helping to develop specifications and resources. For the last few years he was a Subject Advisor and was responsible for the *Maths Emporium* website.

Kathryn Darwin @arithmaticks

Kathryn is a Lead Practitioner of Maths and Teaching at Leeds City Academy. Passionate about teaching for understanding, she has a place as a Secondary Mastery Specialist with NCETM. As an unashamed teaching and maths geek, Kathryn frequently leads #MathsCPDChat and various CPD sessions within school and nationally. When she is not teaching, she can often be found walking Albie the cockapoo, or in the kitchen baking brownies.

Mark Dawes @mdawesmdawes

Mark is a maths teacher and teacher educator based near Cambridge. He is the creator of the *Quibans* and *What the Graph* websites for teachers of Core Maths.

Nathan Day @nathanday314

Nathan is a maths teacher at a secondary academy in Nottingham. He is the creator of *Interwoven Maths* interwovenmaths.com, a site sharing tasks using interweaving - bringing together topics from across mathematics.

Contributors

Dawn Denyer @mrsdenyer

Dawn has been the Head of Department of two large and successful maths departments in the South East and is now an Assistant Headteacher with responsibility for STEM and E-learning. She has a Masters in Mathematics Education, with her dissertation research focusing on students' understanding of ratio and proportion. She is an enthusiastic user of educational technology and her article on 'The role of a Head of Mathematics Department in ensuring ICT provision and use within lessons' is published in *Mathematics Education with Digital Technology* by Adrian Oldknow. Dawn is an accredited NCETM Professional Development Lead and Secondary Mastery Specialist providing training and support through Sussex Maths Hub.

Stella Dudzic @StellaDudzic

Stella taught maths for 22 years in secondary schools, including 9 years as Head of Department, before joining MEI in 2006 where she is Director for Curriculum and Resources. Her work in curriculum development includes designing new qualifications and overseeing the production of teaching resources. In addition to this, she is also a textbook author and editor and regularly leads CPD for teachers.

Sheena Flowers @sheena2907

Sheena has been working as a teacher for over 18 years and has recently taken on a new role leading maths across a trust. She has led workshops at multiple Maths Conferences including conferences with the Mathematical Association and La Salle. She is well known on Twitter for running contentious polls and enjoys building LEGO and playing board games in her spare time.

Dave Gale @reflectivemaths

Dave has been teaching in the South West since 2001. As a Core Maths advocate, he loves creating and sharing resources, especially for estimation and critical analysis topics and leads CPD sessions locally and nationally. When not teaching, he's probably playing *Magic: the Gathering* or trying to improve his juggling.

Contributors

Charlotte Hawthorne @mrshawthorne7

Charlotte is a lead practitioner in North Staffordshire. She posts regularly on Twitter and loves to create resources and write about ideas to teach maths, sharing these on her website, sketchcpd.com. Charlotte is an accredited PD lead, enjoys speaking at conferences and recently has been involved in authoring new KS3 resources with OUP.

Melanie Muldowney @Just_Maths

Mel is a maths teacher and Specialist Leader of Education at a West Midlands School. She previously worked at a school given the accolade "Most Improved School in England" and was part of the team that won TES Maths Team of the Year. She is one third of the team at *JustMaths* (the voice of the blog). Mel has more recently worked with the Kangaroo Maths team to write and publish maths revision guides. In addition she works with Pearson as a Credible Specialist supporting schools and delivering GCSE-specific CPD.

Susan Okereke @DoTheMathsThing

Susan is a maths teacher and communicator. She is a Lead Practitioner in maths at St Mark's Academy in South London and is an Assistant Maths Hub Lead at London South East Plus Maths Hub, leading on the Year 5-8 Continuity project and the Non-Specialists Subject Knowledge for Teaching Mathematics project. She also co-hosts the Maths Appeal Podcast with Bobby Seagull, writes the website dothemathsthing.com and regularly presents at various maths events and projects for students, teachers and the general public. Susan is on a mission to show that maths is everywhere and for everyone!

Chris Pritchard

Chris Pritchard is one of the editors of *Mathematics in School* and the author or editor of eight books published by The Mathematical Association. He is a past Chair of the Scottish Mathematical Council and a past President of The Mathematical Association. He taught secondary mathematics for 39 years in settings ranging from a large urban comprehensive to a tiny remote rural school, much of the time as Head of Department.

Contributors

Rhiannon Rainbow @Noni_Rainbow

Rhiannon is a maths teacher with 20 years' experience teaching secondary. She completed a 4 year BSc Hons QTS Primary Maths EY before making the move to secondary where she has held a variety of roles. In January 2018 she joined the Greenshaw Learning Trust and in September 2020 started her role as the School Improvement Lead (Mathematics) across the Trust. She has recently become a Fellow with the Chartered College of Teaching and is an NCETM Secondary Mastery Specialist and Professional Development Lead. Rhiannon co-founded the GLTBookClub with Dave Tushingham.

Peter Ransom

Peter was a mathematics teacher at state secondary schools for over three decades. He is a Past President and previous Chair of Council of The Mathematical Association and is now fully retired, spending time at his local observatory and playing oboe in a concert band. He has been actively involved with many mathematical organisations and received an MBE for voluntary work in mathematics education in 2019. He feels his teaching was enhanced by the application of mathematics in history, which he did in period costume, both nationally and internationally.

Nikki Rohlfing @heavymetalmaths

Nikki is a heavy metal roadie turned secondary school maths teacher. He has a website called heavymetalmaths.com, where he writes a range of mathematical or musical reviews and thoughts. He is also the creator and host of the online *Can't Get No Mathisfaction* events.

Jennifer Shearman @jenshearman

Dr. Jennifer Shearman is Director for Evaluation at NCETM and recently the MA Education Course Director at Canterbury Christ Church University. Jennifer has led and tutored on numerous routes into teaching: Teach First, School Direct, PGCE and Inspire. Jennifer qualified as a teacher of mathematics through Teach First and taught at schools in London and Kent. Jennifer's Doctorate research utilised Q methodology to explore mathematics teachers' understanding of mastery.

Rob Southern @mrsouthernmaths

Rob teaches Maths and Further Maths at an 11-18 school in Salisbury. He has a YouTube channel called "Maths in an empty classroom".

Contributors

Dave Taylor @taylorda01

Dave is a 15th-year mathematics teacher at Mount St Mary's Catholic High School, Leeds, and Maths CPD Lead at Complete Mathematics. He is best known as the author of the *Increasingly Difficult Questions* maths resources at taylorda01.weebly.com, where you can also find his latest offering - Backward Faded Maths - and is also host of *Teaching Together,* the Complete Mathematics podcast.

Dave Tushingham @DaveTushingham

Dave is a maths teacher with 20 years of experience. Dave qualified as an Advanced Skills Teacher whilst working in Wiltshire before returning to South Gloucestershire to work as an AST and Head of Mathematics. Currently working as a Lead Practitioner within the Greenshaw Learning Trust in Bristol, he has recently become a Fellow with the Chartered College of Teaching and is an accredited Professional Development Lead through the NCETM. Dave is following his passion for continually developing his interpretation of the principles behind Teaching for Mastery in mathematics and co-founded the GLTBookClub with Rhiannon Rainbow.

Susan Whitehouse @Whitehughes

Susan is a teacher of A level Maths and Further Maths, and also delivers professional development to maths teachers. She is co-author of the *Hodder Education A level Maths* textbooks and has developed many teaching resources which she shares via her website susanrwhitehouse.wixsite.com/maths.